PROUD HERITAGE

A PICTORIAL HISTORY OF BRITISH AEROSPACE AIRCRAFT

Compiling Editor: **Phil Coulson**
Design Advisor: **Louise Perry**
Captions: **Christopher Whitehead**

Published by The Royal Air Force Benevolent Fund Enterprises,
RAF Fairford, Glos GL7 4DL.

ISBN 1-899808-10-8

THE BRITISH AEROSPACE STORY

The history of British Aerospace is in many respects the history of aviation progress in Britain. Any deliberate attempt to bring into being one of the world's leading aerospace companies from origins as humble and diverse as the real ones, would have been looked upon as folly. And yet, individual circumstances combined with global influences such as the two World Wars to forge the links, and consolidate the expertise, that eventually brought about, in 1977, the formation of British Aerospace. At that time, the elements that merged to form the new company were themselves well-established as leaders in the aviation field. The British Aircraft Corporation, Hawker Siddeley Aviation, Hawker Siddeley Dynamics and Scottish Aviation were well-founded and successful, and the aircraft they produced bore company names that went right back to the roots of British aviation.

ALLIOTT VERDON-ROE

So where did it all begin? It might be logical to take as the foundation stone the first powered flight by an Englishman. If so, the honour goes to Manchester-born Alliott Verdon-Roe who, on 8 June 1908, got airborne from the Brooklands motor-racing track at Weybridge, Surrey, and flew for a distance of about 100 feet. That historic flight was the first of several made that day but, curiously, none of the flights was considered worthy of recognition and Alliott Verdon-Roe, later Sir Alliott, was never officially credited with being the first Englishman to fly. He did, however, indisputably become the first British man to fly over British soil in a British aeroplane powered by a British engine. This was on 13 July 1909 when his "Avroplane", a triplane powered by a 9hp JAP engine, took off from Lea Marshes near London.

A few months later, in February 1910, Verdon-Roe joined forces with his brother, Humphrey, to build aeroplanes in a factory that Humphrey owned in Brownfield Mill, Manchester. That enterprise flourished and, in January 1913, A V Roe & Co Ltd was registered as a public company. The first of the legendary British aircraft names emerged from the Manchester factory when, in 1913, the Avro 504 first took to the skies. The Avro 504 became the basic trainer for the newly-formed Royal Flying Corps and, later, for the RAF who trained untold thousands of pilots on it until 1932.

JOHN SIDDELEY

But perhaps the foundation stone of British Aerospace should be set in events prior to Verdon-Roe's historic flight, great achievement though that was. Few would wish to dispute that British Aerospace exists today because of the people involved in its history, those whose vision, courage, inventiveness and pioneering spirit made happen the milestone events. The birth of aviation brought together strands of "the right stuff" that were already making their mark in other fields. Amongst such was the young John Siddeley who, after entering the history books as the first man to cycle from Land's End to John o' Groats, progressed to motor engineering and formed, in 1902, the Siddeley Autocar Co to produce a four-cylinder car of his own design. This venture was a success, and Siddeley went on to form the Siddeley Deasy Motor Car Co at Parkside, Coventry. Although not involved in aircraft production until after the outbreak of WW1, Siddeley became a great influence in the developing British aircraft industry, of which more later.

ROBERT BLACKBURN

Soon after Verdon-Roe's tentative victory over gravity, other young men whose imaginations had been caught by the prospect of flight raced to become the next to fly. Robert Blackburn, who had spent the previous year busily building his own flying machine, achieved his first powered flight from a beach in Yorkshire in the spring of 1909. His aircraft was underpowered and when, in early 1910, he attempted to make a turn, it crashed and was badly damaged. Undaunted, Blackburn set about building another aircraft, the design of which he based on the successful French Antoinette monoplane. The Blackburn monoplane had an uneventful first flight, piloted by B C Hucks, from a beach near Filey on 8 March 1911. Even before his Antoinette-based monoplane had flown, Blackburn was working on his third aircraft, the Mercury. The aircraft was completed in time to be exhibited at the 1911 Olympia Aero Show, and the banner above it proclaimed it to be the product of The Blackburn Aeroplane Company. In the same year the apparently indefatigable Blackburn set up a flying school, and a factory, in some disused stables in Leeds. The need for expansion early in 1914 caused a move of the Blackburn facilities to the Olympia roller-skating rink in Roundhay Road, Leeds, with an adjacent aerodrome in Roundhay Park. This location served The Blackburn Aeroplane and Motor Co Ltd until 1929, when the company moved to a new factory and aerodrome at Brough on the banks of the Humber.

GEOFFREY de HAVILLAND

At the same time as Robert Blackburn was exploring the technicalities of flight, the young Geoffrey de Havilland was constructing his first flying machine. Notwithstanding his

The Antoinette was the inspiration for Blackburn's second monoplane. **(SSPL)**

A Farman aeroplane at Brooklands looks only slightly more substantial than Eardley Billings' Pilot Training Apparatus in the foreground. Geoffrey de Havilland adopted the Farman design after his first aircraft crashed. **(SSPL)**

meticulous attention to detail, a quality that served him well throughout a lifetime in aviation, his biplane broke up as soon as it left the ground on its first flight in December 1909. So severe was the crash that the only item salvaged was the engine, which had been designed and built by de Havilland. Whereas Blackburn turned to the Antoinette for the basic design of his second aircraft, de Havilland favoured the biplane configuration used successfully by Frenchman Maurice Farman. de Havilland first piloted his new biplane from a meadow near Newbury, Berkshire, in September 1910 and went on to complete many more flights in which the concept, if not the name, of "exploring the flight envelope" became established.

The expense of his flying activities left Geoffrey de Havilland short of money, and he was therefore glad to be offered, in December 1910, the post of designer and test pilot at the newly-formed Royal Aircraft Factory at Farnborough. His biplane was bought by the Factory for £400, and subsequently became the FE1, the first of an illustrious line of Factory designs. In May 1914 de Havilland was persuaded to join The Aircraft Manufacturing Co Ltd (known as "Airco") which had been established by G Holt Thomas near Hendon Aerodrome. In his post of Chief Designer, Geoffrey de Havilland was most influential, but the post-War slump in aircraft production forced Holt Thomas to sell Airco to the Birmingham Small Arms organisation, and de Havilland was out of work. Not the sort of man to give up, de Havilland mustered all the cash that he had and, with generous support from Holt Thomas, started The de Havilland Aircraft Co Ltd on 25 September 1920. As the new company established itself at Stag Lane, near Hendon, de Havilland became increasingly frustrated with the Air Ministry and its contract and specification procedures. He decided to concentrate on the production of civil aircraft, and it was the Moth series of light aircraft which served to put the company in a sound financial position. Another of the early pioneers to be knighted, Sir Geoffrey's chosen initials, DH, were associated with a range of designs from the DH1 to the DH125.

THOMAS SOPWITH

Thomas O M Sopwith gained airborne experience sooner than most of his fellow aviation pioneers, but his flights were in a balloon. He made several such flights in 1908 and 1909 but, because he was under 21, was refused a balloon pilot's certificate. Sopwith, a keen motorist and yachtsman, soon turned his attention to powered flight and, in 1910, bought a Howard Wright monoplane and taught himself to fly. Within a month, on 21 November 1910, he gained his Royal Aero Club aviator's certificate (No 31), having already broken, eleven days earlier, two British records by flying a distance of 107 miles in 3 hours 12 minutes.

Just as Geoffrey de Havilland had applied himself to the technicalities of the flying machine, Thomas Sopwith applied himself to the development of flying as a skill. His reputation as a pilot was enhanced when, in 1911, he won a number of flying competitions in America, and in 1912 he opened a flying school at Brooklands. One of Sopwith's pupils at the Brooklands school was a certain Major Trenchard who, as Major General Trenchard, six years later founded the Royal Air Force.

Later in 1912 Sopwith (later Sir Thomas) decided to build aeroplanes, and for this purpose bought a disused roller-skating rink in Kingston-upon-Thames. Sopwith's contribution to those early days of aviation was profound, and who can judge the value of that contribution to the outcome of the aerial war that was soon to take place? Undoubtedly, though, some of the success of The Sopwith Aviation Co, registered in 1913, was due to the recruitment a year earlier of a young Australian mechanic, Harry Hawker.

HARRY HAWKER

Thomas Sopwith taught Harry Hawker to fly in 1912. In Sopwith's own words "The pupil became the master of the art", and Hawker soon became the chief test pilot for The Sopwith Aviation Co. All through the War years Hawker honed his skills, and the excellence of the Sopwith stable owed much to his intuitive command of flying. The vacuum in aircraft production at the end of the Great War saw many manufacturers in difficulty. Sopwith chose to put The Sopwith Aviation Co into voluntary liquidation in September 1920, but in December of the same year formed a new company, H G Hawker Engineering Co Ltd, named after his test pilot whose name was almost as well known as that of Sopwith himself. Hawker's fame had been enhanced by an attempt, in May 1919, to achieve the first non-stop crossing of the Atlantic, and his subsequent almost-miraculous rescue when the attempt failed. Sadly, Harry Hawker did not live long to serve the new company. On 12 July 1921 he was killed near Hendon when his Nieuport Goshawk crashed while practising for the Aerial Derby.

NOEL PEMBERTON-BILLING

Although his own name never achieved the legendary status given to those of many of his peers, Pemberton-Billing coined a name that is secure in aviation history. A wealthy and enterprising man, well-known for his exploits as a motorist and yachtsman, Pemberton-Billing learned to fly in 1913. For a wager of £500, he learned to fly and gained his aviator's certificate not only in one day, but before breakfast! Nobody would hire him an aeroplane for

Harry Hawker standing beside his Sopwith Bee single-seat biplane. (IWM)

this escapade, so he bought a Farman biplane and hired an instructor, Barnwell, from the Vickers flying school at Brooklands. With his nautical background, and a boat factory at Woolston on Southampton Water, Pemberton-Billing's stated aim was "to build boats that could fly, rather than aeroplanes that could float". When he registered the firm of Pemberton-Billing Ltd in 1914, he chose as the factory's telegraphic address the name 'Supermarine', although the company's first flying boat, a tractor biplane, was shown at the 1914 Olympia Aero Show under the unimaginative name of 'PB1'. No record exists that the PB1 ever flew, and Pemberton-Billing joined the Royal Naval Air Service in 1916. He sold his interest in the company which then changed its name to The Supermarine Aviation Works Ltd.

SIR GEORGE WHITE

The emphasis now moves away from the influence of pioneering aviators who founded companies in pursuit of their aviation dreams, to those businessmen who saw in aviation an opportunity to make money. One such was Sir George White, the pioneer of the electric tramway system in his native city, Bristol. A flying display convinced Sir George that aviation had a future, and he formed The British and Colonial Aeroplane Co Ltd on 19 February 1910. At the same time, Sir George registered three other companies, one of which was The Bristol Aeroplane Co.

The British and Colonial Aeroplane Co. started operations in 1910 at Filton, four miles north of Bristol, in two large sheds hired from Sir George's tramway company. The idea of building French Zodiac biplanes under license was abandoned when it was discovered that the design failed to fly. A subsequent venture, based on the Farman biplane, resulted in the famous Bristol Boxkite which proved to be an ideal trainer. The company formed two flying schools, one at Larkhill on Salisbury Plain and the other at Brooklands, where it used the Boxkite to train more than 300 pupils by August 1914. The Boxkite was the first commercial export success, with more than 20 of the 76 Boxkites built sold abroad.

By the end of the war the company had acquired an additional site at Brislington, a suburb of Bristol, and the imposition of Excess Profits Tax (the principal cause of many smaller businesses collapsing after the war) led to The British and Colonial Aeroplane Co being taken over by The Bristol Aeroplane Co in 1920.

THE GLOUCESTERSHIRE AIRCRAFT CO LTD

The need to subcontract aviation engineering work during WW1 provided a development opportunity for H H Martyn & Co Ltd of Cheltenham, a firm with a good reputation for high-class woodwork. The company undertook work for Holt Thomas of Airco, and he was so impressed that he proposed a partnership in a separate aviation company. So it was that in June 1917 The Gloucestershire Aircraft Co Ltd was formed.

By 1918 the company was hard at work building Bristol Fighters and Nieuport Nighthawk single-seat fighters, a product of The Nieuport and General Aircraft Co. Nieuport was forced to close after the war and The Gloucestershire Aircraft Co, with considerable foresight, bought not only a quantity of Nighthawk components but also recruited Nieuport's chief designer, Henry P Folland. Folland, like de Havilland, had been a designer at Farnborough where he was largely responsible for the SE5 which, with the Sopwith Camel, had vied as the RFC's most successful fighter.

VICKERS SONS & MAXIM

Vickers Sons & Maxim were well-established in the armament business when they were asked by the Admiralty, in 1908, to build a rigid airship on the lines of the German Zeppelin. In spite of a total lack of experience in building airships, Vickers commenced construction of an airship, unofficially known as the Mayfly, in 1909. There were long delays and much wrangling with the

Admiralty over funding, and the name 'Mayfly' must surely have attracted some ribald comment! In the event, 'Mayfly' became 'Neverfly' because, shortly after completion in 1911, the airship broke its back without ever having flown.

Notwithstanding the Mayfly failure, Vickers formed an aviation department in 1911 and acquired a licence from the French firm Robert Esnault Pelterie to build the REP monoplane. The first of these was purchased in France, modified by Vickers, and tested at Joyce Green aerodrome near Dartford, Kent. Also in 1911, the company adopted the shortened title of Vickers Ltd and opened a flying school at Brooklands to where, in 1915, they moved their aircraft factory from its original location at Erith, Kent.

After building several variants of the REP monoplane, Vickers produced a series of pusher aircraft, the most famous of which was the FB5. Known as the Vickers Gunbus, the FB5 was Britain's first practical fighting aeroplane and gave good account of itself during the opening months of the First World War.

So rapid was the growth of Vickers' aviation engineering knowledge, that the Vimy was designed, built and tested all within five months. Too late to prove its worth as a bomber in WW1, the Vimy nevertheless achieved fame when, flown by Alcock and Brown, it became the first aircraft to fly non-stop across the Atlantic. The Vickers Vimy was also the first aircraft to fly from England to Australia.

The Vickers Vimy was the first aircraft to fly non-stop across the Atlantic, and the first to fly from England to Australia. **(IWM)**

SIR W G ARMSTRONG WHITWORTH & CO LTD

The great rival to Vickers Sons & Maxim in the armament field, the firm of Sir W G Armstrong Whitworth & Co Ltd decided to form an 'aerial department' after approaches made by the Admiralty and the War Office. The Admiralty was inviting tenders for the construction of rigid and non-rigid airships, and the War Office wanted Armstrong Whitworth to build aeroplanes. Land was purchased at Barlow, Yorkshire, for the building of an airship factory, and an aeroplane production facility was installed in one of the company's existing buildings at Scotswood in Newcastle-upon-Tyne. It was here that the first batch of Royal Aircraft Factory-designed BE2s was built for the War Office in 1913. Later that year, Armstrong Whitworth acquired a disused roller-skating rink (quite a habit, this, for aircraft companies!) at Gosforth, north of Newcastle, and an aerodrome was established on Newcastle's Town Moor. Armstrong Whitworth contributed four types of aircraft to the 1914-18 war, the most successful of which was the FK8 biplane used for artillery observation and photography. The abrupt cancellation of orders at the end of the war forced the closure of Armstrong Whitworth's aerial department in October 1919, but this is where John Siddeley reappears!

INTER-WAR AMALGAMATIONS

ARMSTRONG WHITWORTH AND SIDDELEY DEASY

The outbreak of war in 1914 had an enormous effect on the fledgling aviation industry in Britain. In the first ten months of the war, up to May 1915, 530 aircraft were built in British factories. During the last ten months of the war, that figure had become 26,685! By far the majority of those aircraft were built by subcontractors who, prior to the war, had no experience at all of aviation engineering. When the war ended, and various austerity packages were introduced, many small companies went to the wall whilst others flourished. The twin blows of cancelled orders and the Excess Profits Tax made those companies who were serious about a future in aviation look for strength in partnership.

At the time of Armstrong Whitworth's closure of its aerial department, John Siddeley had already persuaded Armstrong Whitworth to buy the Siddeley Deasy company which had been producing aircraft as well as aero engines. An exciting new prototype, the Siddeley Siskin, made the offer attractive and, late in 1918, The Armstrong Whitworth Development Co Ltd was formed. This subsidiary company itself had two subsidiaries, Armstrong Siddeley Motors Ltd and Sir W G Armstrong Whitworth Aircraft Co Ltd. These all had their headquarters in Coventry.

Under Siddeley's capable leadership, the Coventry-based business flourished in spite of the hard times. The Newcastle element of Armstrong Whitworth was languishing, however, and Siddeley made the courageous decision to offer £1,500,000 for the whole Coventry operation. The deal was concluded in December 1926 and, the following year, the name of the holding company was changed to The Armstrong Siddeley Development Co Ltd with its two subsidiaries retaining their names. The fortunes of the Newcastle-based Armstrong Whitworth company continued to decline until, in 1927, the long rivalry with Vickers was resolved and the two armament giants merged to become Vickers-Armstrong Ltd.

AVRO AND CROSSLEY

In 1920, the Avro company became a wholly-owned subsidiary of the Crossley Motor Group, albeit retaining its own name, and in 1927 the Crossley Group sold Avro, again under its own name, to John Siddeley. Sir Alliott Verdon-Roe severed his connections with the company that bore his name and took up a controlling interest in the firm of S E Saunders who built boats and flying boats in the

Isle of Wight. That company became known as Saunders-Roe Ltd and, in 1927, became a partly-owned subsidiary of de Havilland until 1959 when it was sold to the Westland Aircraft Co.

SUPERMARINE AND VICKERS

The Supermarine Aviation Works Ltd was prospering in Southampton and, in 1920, appointed as its chief designer Reginald Mitchell, who had joined the company in 1917. R J Mitchell's designs enjoyed outstanding success in the prestigious Schneider Trophy races, and culminated in the immortal Supermarine Spitfire, first ordered in 1936. Meanwhile, Vickers had turned its aviation department into a separate subsidiary called Vickers (Aviation) Ltd, which bought Supermarine in 1928. No immediate changes were made and the Southampton team was intact when the company name changed to The Supermarine Aviation Works (Vickers) Ltd in 1931. In 1938, however, both subsidiaries were subsumed into Vickers-Armstrong Ltd.

HAWKER, GLOSTER AND SIDDELEY

The final significant merger in the inter-war years occurred in 1935 when Sir John Siddeley decided to sell his companies to Hawkers. At the converted roller-skating rink at Kingston-upon-Thames, H G Hawker Engineering Co Ltd had recruited a young draughtsman called Sydney Camm who, in 1925 when the post became vacant, took over as chief designer. Camm was a designer of exceptional ability who did much to develop aircraft technology in the inter-war years. In 1933 the company adopted the simpler name of Hawker Aircraft Ltd and, soon after, Camm's expertise produced the Hurricane which attracted an initial order of 600 aircraft.

The Gloucestershire Aircraft Company had not been idle, and its salesmen were bringing in good orders from overseas for its Folland-designed fighters. Tired of the constant mis-spelling of their company's rather cumbersome name, the salesmen convinced their directors to adopt a more simple title and, in 1926, the company became Gloster Aircraft Ltd. The Gloster Grebe and Gamecock fighters were successful but, in the early thirties, business was declining and a takeover of Gloster by Hawkers in 1934 was accepted. Henry Folland resigned as chief designer in 1937, and his place was taken by George Carter who went on to design Britain's first jet aircraft, the Gloster E28/39, which flew in May 1941.

The company that resulted from Hawkers acquisition of the Siddeley companies and Gloster was initially called the Hawker Siddeley Aircraft Co, but soon became the Hawker Siddeley Group including Hawker Aircraft, Armstrong Whitworth Aircraft, Armstrong Siddeley Motors, Avro and Gloster.

NEW COMPANIES APPEAR

AIRSPEED LTD

While the established names in aviation were jockeying for position, new companies appeared from time to time. Some of those companies, now largely forgotten, played a part in the development of British Aerospace. Airspeed Ltd was formed in 1931 but was underfunded and struggled to survive. In 1934 the company was rescued by shipbuilders Swan Hunter who formed a new company called Airspeed (1934) Ltd. Now properly resourced, Airspeed gained a foothold and, within three years, had designed and built the Oxford which became widely used as an advanced trainer by the Commonwealth air forces. Airspeed also built the Horsa troop-carrying glider and, during the Second World War, carried out a considerable amount of sub-contract work for de Havilland. This close association led to Airspeed being bought by de Havilland in 1940, although it continued in a semi-autonomous role until, in 1951, it became the Airspeed Division of de Havilland.

GENERAL AIRCRAFT LTD

Another small company, General Aircraft Ltd, formed in 1931 and established itself at Hanworth Aerodrome, Middlesex. It concentrated mainly on sub-contract work but, during the war, designed and built the Hamilcar tank-carrying glider. After the war the company merged with Blackburn to become Blackburn and General Aircraft Ltd. The GAL 60 Universal Freighter, whose freight floor design was influenced by Hamilcar technology, became the Blackburn Beverley and entered RAF service in 1956.

THE PERCIVAL AIRCRAFT COMPANY

In 1932 Edgar Percival formed a company to market the Vega Gull monoplane which was proving popular with private owners. The Percival Aircraft Co established a factory adjoining Luton Aerodrome and won a lucrative order from the Air Ministry for a military version of the Vega Gull. Before long, large numbers of Percival Proctor aircraft were leaving the factory for use as radio and navigational trainers with the RAF and the Royal Navy. Percival's first twin-engined aircraft, the Petrel, had a promising start before the outbreak of war in 1939 caused production to cease. In 1944 the company became part of the Hunting Group which, in 1954, changed its title to Hunting Percival Aircraft Ltd, later to change again, in 1957, to Hunting Aircraft Ltd.

FOLLAND AIRCRAFT LTD

The British Marine Aircraft Company Ltd was formed at Southampton in 1935, but underwent reorganisation in 1937 under the managing directorship of H P Folland to become Folland Aircraft Ltd. The company was busy throughout the war on sub-contract work and, after some adjustment in the late forties, began work in 1951 on a private venture light fighter, the Midge. That design developed into the Folland Gnat which became the advanced jet trainer for the RAF, and the mount of the famous Red Arrows aerobatic team.

THE ENGLISH ELECTRIC COMPANY LTD

Although English Electric became known in 1938 as a major producer of Handley Page bombers, the company's genesis pre-dated the First World War. The Coventry Ordnance Works, founded in 1912, had produced some aircraft of their own design before concentrating on the construction of other companies' aircraft during the war. Another company, The Phoenix Dynamo Manufacturing Company of Bradford, had experience of building flying boats for the government and, in 1918, these two companies amalgamated. Before the amalgamation was finalised, three other companies, Dick Kerr & Co of Preston, Willans & Robinson Ltd of Rugby, and the Siemens Dynamo Works of Stafford, joined forces to form the English Electric Co Ltd. The new company went on to produce successful prototypes, mainly flying boats, but with no production orders forthcoming decided, as a temporary measure, to close its aircraft department in 1926.

The 'temporary' closure lasted until 1938 when the contracts were placed to build Handley Page bombers, an activity which occupied English Electric throughout the war. After the war, the company set up its own design organisation under the direction of W E W 'Teddy' Petter, and produced two classic military aircraft, the Canberra bomber and the Lightning, both of which types saw more than 25 years of front-line service with the RAF. The aircraft division of the English Electric Co became a wholly-owned subsidiary of the parent company in 1959, under the name English Electric Aviation Ltd.

SCOTTISH AVIATION LTD

The final major piece of the British Aerospace mosaic was brought into being by two intrepid aviators who, in April 1933, proved that the pioneering spirit of aviation's early days lived on. The Marquess of Clydesdale (later the Duke of Hamilton) and fellow-Scotsman David McIntyre decided to be the first men to fly over the summit of Everest. They succeeded at their first attempt, not without peril, and on return to their native land established a flying school at Prestwick, on the west coast of Scotland. Unlike other industries, the aircraft repair industry was not unduly hampered by the lack of industrial infrastructure because, of course, aircraft could be flown in and out of more remote areas. Thus the company formed by the Marquess and McIntyre, Scottish Aviation, prospered by undertaking overhaul and repair work for the burgeoning aircraft industry.

The success of Scottish Aviation, in turn, led to the development of Prestwick Aerodrome, later to become the essential terminal of the wartime Atlantic ferry organisation. The flying school continued until 1941, by which time Scottish Aviation had moved into larger premises and was working to capacity in the modification for RAF service of imported US aircraft, and the repair of other combat aircraft.

After the war, Scottish Aviation designed and built the Pioneer and the Twin Pioneer, both of which types saw RAF service. As production of those aircraft ceased, Scottish Aviation took over production of the Jetstream aircraft from the ailing Handley Page company, and also took over production of the Bulldog from Beagle. Scottish Aviation was acquired by the Cammell Laird Group in 1966.

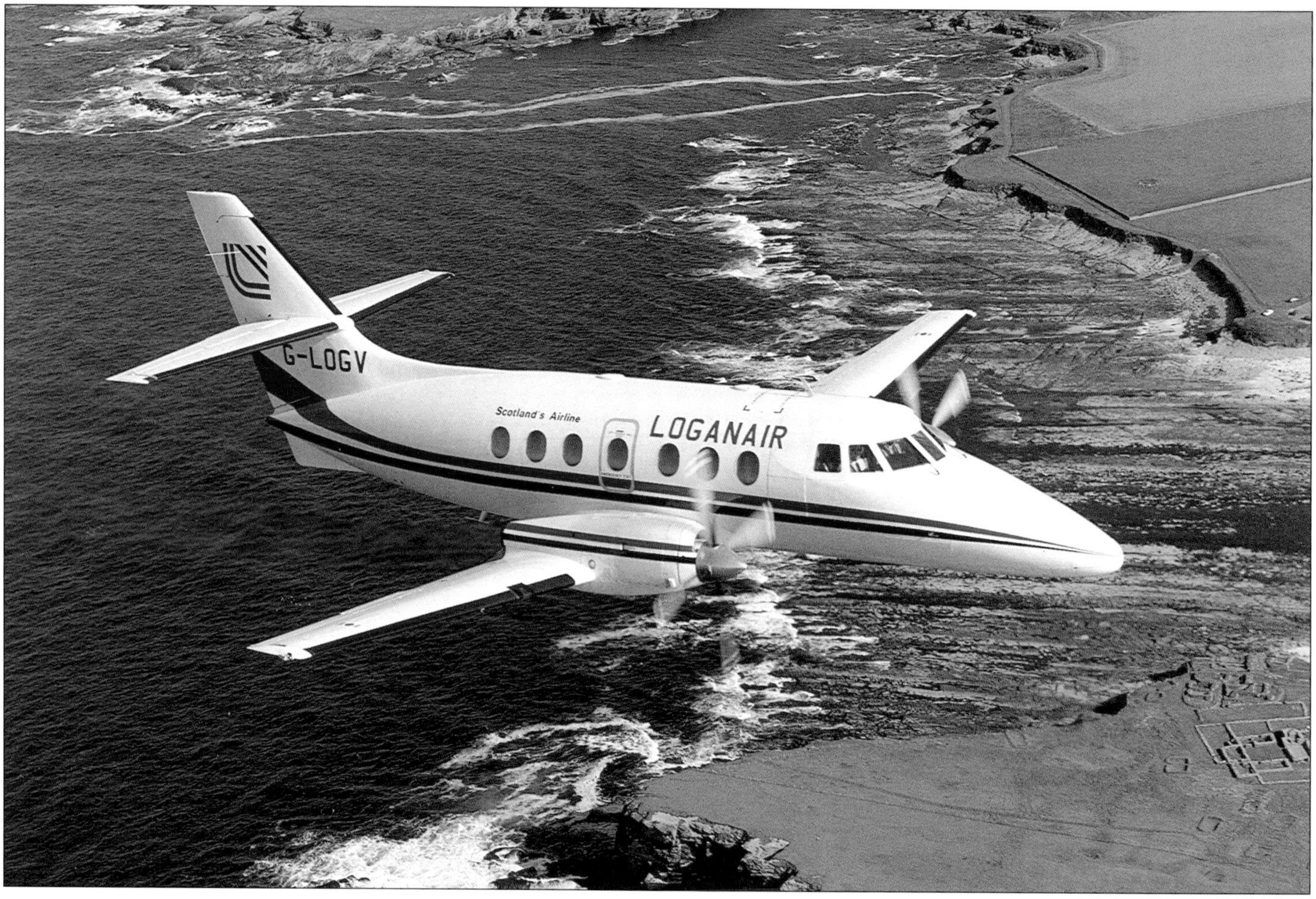

Scottish Aviation took over production of the Jetstream from Handley Page. Seen here in the livery of Scotland's Loganair, the Jetstream has also sold well in the US where it is a popular commuter passenger aircraft. (BAe)

POST-WAR CONSOLIDATION

The end of the Second World War brought with it tremendous new challenges and opportunities for the aviation industry. The pace of development had been forced by military requirements and, arguably, more progress had been made in aircraft design technology, component manufacture and systems reliability than would have been made in decades of peacetime. Other wartime developments had helped to change aviation for ever. The improvements in navigation systems, both airborne and on the ground, increased greatly the safety of flying, as did the advent of radar technology and radio communications. There now existed a large cadre of highly trained and experienced aircrew, groundcrew and design staff, all eager to put their hard-won skills to peacetime use.

But the most significant product of wartime aircraft development had yet to be exploited. The invention of the gas-turbine engine was as significant an event, almost, as the Wright brothers' first powered flight. Piston engine technology had made great strides, but the jet engine was to revolutionise aviation. If the dream of global peace had materialised then, perhaps, jet engine development would have been a drawn-out affair. International tensions continued, however, and the air forces of the world began to look for aircraft that were significantly more capable than those that had been used in the Second World War. Britain was at the forefront of jet technology, and well-placed to meet the new demand with such types as the Meteor, Vampire and Hunter which equipped the RAF and other air forces. The emergence of the so-called Super Powers, and their respective spheres of economic influence, led to rearmament based on political lines rather than on the relative merits of the equipment available. Nevertheless, the adoption by the US forces of the Canberra and, in later years, the Harrier and Hawk, demonstrated that although political and economic muscle had denied Britain the global market she might otherwise have enjoyed, the technical excellence of British aircraft was undiminished.

In the field of civil aviation, progress was every bit as swift as with military aircraft. Initially, the surplus wartime aircraft were hastily converted, none too cosmetically in some cases, to provide the capacity for foreign travel that government, business and the rich had come to expect. The conversion of Lancasters into Lancastrians, and Halifaxes into Haltons was superseded by development of aircraft such as the York and Tudor from wartime designs. Again, though, the death knell for these faithful piston-engined types was being sounded by the roar of the jet engine, and when the Comet appeared in the early '50s there was no doubt as to where the future lay for civil aviation.

The American aviation industry was equally advanced by now, and the appearance of the Boeing 707, Douglas DC-8 and other jet-powered aircraft was a serious challenge to British aircraft manufacturers. Still reeling throughout the '50s from the cost of the war, Britain simply did not have the capacity to challenge the US industrial giants, but the vision and skill of her designers were undiminished. Despite leading the way with the world's first turbo-prop airliner, the Viscount, and inaugurating the first trans-Atlantic regular jet service with the Comet, British aviation industry required a strategy that would ensure its survival in an increasingly competitive world.

Thus, in 1960, the Hawker Siddeley Group which already comprised Armstrong Whitworth, Avro, Gloster and Hawkers, in accordance with government policy acquired the Blackburn, Folland and de Havilland companies. Further rationalisation was completed in the same year when Bristol Aircraft, English Electric Aviation, Vickers Armstrong (Aircraft) Ltd and Hunting Aircraft were brought together under the title of the British Aircraft Corporation. In 1963 the Hawker Siddeley Group was divided into two companies, Hawker Siddeley Aviation and Hawker Siddeley Dynamics but, all the time, the strategic eye was on the eventual merging of these two companies, and BAC, under one name.

The next ten years were highly productive, with several new types coming off the production line for both civil and military use. The graceful Vickers VC10, Trident, Harrier, HS125 and Nimrod, all British designed and built, appeared in the 1960s but, as that decade drew to a close, a new aspect of British aviation strategy was being developed. 1969 saw the first flight of the Jaguar, built as a joint project between BAC and Dassault-Breguet of France, and also the first flight of the supersonic Concorde, developed jointly with Aerospatiale of France. Discussions were well underway with a European consortium to build the Airbus, the A300 version of which first flew in 1972 with wings built in Britain.

No longer could the enormous expense of research and development of a new aircraft type be borne by a single company, or recovered by selling that product to a single market. Collaboration, joint venture, consortium, all became part of the aviation industry language throughout Europe during the early '70s, and the imperative to spread development costs and widen markets made further consolidation of the British aviation industry inevitable.

BRITISH AEROSPACE

THE STRUCTURE

On 17 March 1977 the Royal Assent was given to a bill that had been the subject of long debate in Parliament. The Aircraft and Shipbuilding Industries Bill proposed, among other things, that Hawker Siddeley Aviation, Hawker Siddeley Dynamics, the British Aircraft Corporation and Scottish Aviation be merged to form one new state-owned company whose name should be British Aerospace. Lord Beswick, for some time the chairman of the organising committee, was confirmed as chairman of the new company which came into being formally on 29 April 1977. For the next three years he guided the practical rationalisation of different workforces, factories and projects and, in 1978, finalised British Aerospace's joining of the European consortium Airbus Industrie as a 20% partner with the role of wing-maker. Lord Beswick handed over control, on 22 March 1980, to Sir Austin Pearce whose initial task was to oversee the flotation on the Stock Exchange, on 1 January 1981, of British Aerospace plc. It was not until 1985 that British Aerospace was fully privatised, the British Government selling all its remaining shares in the company with the exception of a special £1 share, necessary to ensure that BAe remains under UK control. The intervening years had seen the start of an ambitious acquisition programme, the aim of which was to broaden the company base and, in 1982, the establishment of the Sowerby

Research Centre at Filton as BAe's scientific research organisation. By 1987 BAe was established as Europe's largest aerospace company, with the widest range of products of any such company anywhere in the world. Within ten years of the consolidation of Britain's aircraft industry, a potent organisation had emerged, well-founded, broad-based, modern and dynamic, capable not only of surviving in the world market but of restoring Britain's lead.

The expansion of British Aerospace plc resulted, in January 1989, in a restructuring programme whereby British Aerospace plc became the HQ management organisation controlling wholly-owned subsidiaries, each marketing its own specialised products under its own name. The previous two years alone had seen the acquisition of Royal Ordnance plc, Steinheil Optronik GmbH, the Ballast Nedam Group, and the Rover Group plc.

For the next two years each of the subsidiaries responded to the challenge of semi-autonomous operation, and some made acquisitions in their own right. A firm foothold was gained in the growing field of satellite technology and, in March 1991, the German small arms and general engineering company, Heckler and Koch GmbH, was acquired.

A significant development in June 1991 was the formation in Rome of Euroflag srl, a European joint venture company to design, develop and build a new medium-lift military transport aircraft that would meet the requirements of the world's air forces for the 21st century.

By the beginning of 1992, continued growth and development necessitated a further rationalisation of British Aerospace. The three previously separate defence companies, British Aerospace (Military Aircraft) Ltd, British Aerospace (Dynamics) Ltd and Royal Ordnance plc, became Divisions of the newly-created British Aerospace Defence Ltd, together with a fourth Division, Systems and Services.

The following month, February 1992, saw restructuring applied to the civil aircraft side of British Aerospace. In this case, three companies emerged from one as British Aerospace (Commercial Aircraft) Ltd divided into British Aerospace Airbus Ltd, British Aerospace Regional Aircraft Ltd, and British Aerospace Corporate Jets Ltd (renamed, in May 1992, Corporate Jets Ltd).

Although 1992 saw the last major changes to the structure of British Aerospace plc, an essential feature of the company is its ability to move with the times, acquiring or selling facilities as required. Thus, in June 1992, the Communications Division of British Aerospace Space Systems Ltd was consolidated with Bishopsgate Systems Ltd, Satellite Management International Ltd, and Starbird Satellite Services Ltd, to form a separate trading company, British Aerospace Communications Ltd.

Assembly of the ATP aircraft was transferred to Prestwick in October 1992, and the type became part of the Jetstream family of regional turboprop aircraft. At the same time, a separate company, Jetstream Aircraft Limited, was formed.

Another joint venture emerged in October 1993 when British Aerospace (Dynamics) and GEC-Marconi formed a company, UKAMS Ltd, to manage and develop their involvement in the Principal Anti-Air Missile System (PAAMS) guided weapons project for the Royal Navy. March 1994 saw the sale of the Rover Group to BMW AG.

COLLABORATION AND JOINT VENTURE

From the principles of collaboration, risk-sharing and joint development practised on a national basis years ago, British Aerospace now has partners across the world. Expansion of these principles first encompassed Europe, resulting in the formation of SEPECAT by BAC and Dassault-Breguet to produce the Jaguar aircraft, still in service with the British, French and Indian air forces. Concorde, always in a class of its own, was built jointly with Aerospatiale of France and began its supersonic passenger services in 1976.

PANAVIA AIRCRAFT GmbH

The momentum for cooperation was increasing and, for the ambitious Tornado project, the tri-national company Panavia was formed by Britain, Germany and Italy. Whilst the Anglo-French Jaguar and Concorde programmes were important, the Tornado project was huge. The problems involved in bringing together elements of the national aircraft manufacturers of three nations, each with its own determined identity, methods and language, at times seemed insurmountable, but the success of the Tornado programme laid the foundations for future collaborative projects. Just under 1000 Tornados, of three variants, have been delivered to the air forces of Britain, Germany, Italy and Saudi Arabia. The German and Italian navies also have Tornados in service. The 72 Tornados for Saudi Arabia were part of Britain's biggest-ever export order.

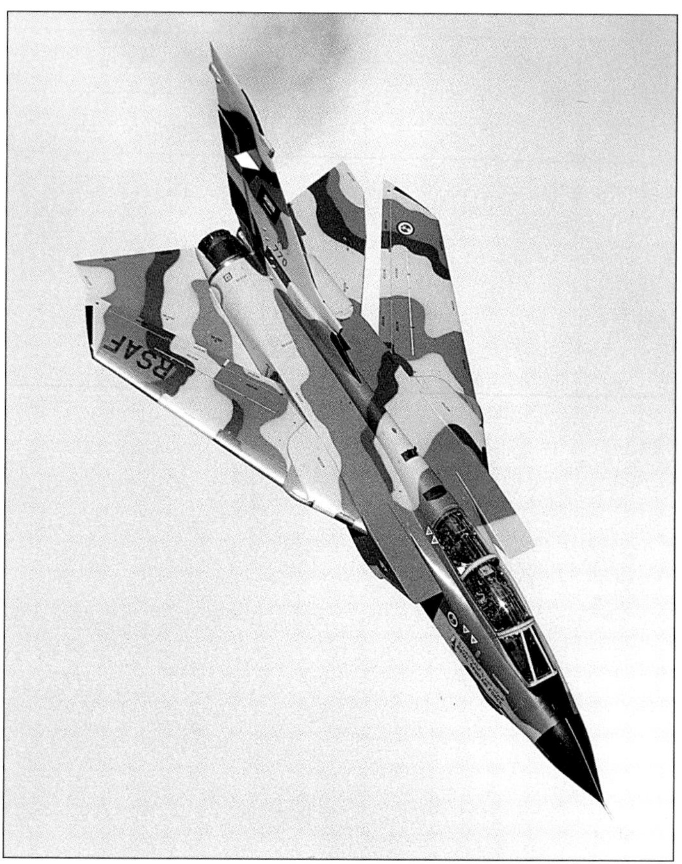

A Panavia Tornado in the markings of the Royal Saudi Air Force. This aircraft was part of Britain's biggest ever export order. (BAe)

AIRBUS INDUSTRIE

The successful Airbus Industrie partnership, launched in 1978, is a triumph of collaboration between British Aerospace, Aerospatiale of France, CASA of Spain, and Deutsche Aerospace Airbus of Germany. Airbus Industrie is now established as one of the world's two largest commercial aircraft manufacturers. British Aerospace Airbus Ltd designs the wings, and builds the primary structure of all Airbus wings. Additionally, BAe builds a fuselage section for the A321, besides other components for the Airbus series. The Airbus family is set for a long life, and British Aerospace is firmly committed to its success.

McDONNELL DOUGLAS

An historic agreement in 1981 marked the beginning of an exciting collaboration with McDonnell Douglas. The US Navy selected the Hawk for its undergraduate jet training requirement, and that led to a joint programme with McDonnell Douglas to produce nearly 300 aircraft, to be called Goshawk T-45A, for delivery in the 1990s.

In a separate agreement, the US Marines opted for the Harrier II Plus to augment its fleet of AV-8Bs which will be upgraded to the new II Plus standard. Ordered by the Spanish and Italian navies as well, the Harrier II Plus will be produced in partnership with McDonnell Douglas in the largest ever Anglo-US aircraft project of all time.

Future collaboration between British Aerospace and McDonnell Douglas involves the development of the Advanced Short Take-off and Vertical Landing (ASTOVL) technology demonstration programme for a new breed of strike/fighter aircraft.

EUROFIGHTER JAGDFLUGZEUG GmbH

A vital step in securing one of Europe's major military aircraft programmes was taken in 1985 when agreement was reached between British Aerospace, DASA of Germany, Alenia of Italy and CASA of Spain to design, develop and manufacture a new European Fighter Aircraft (EFA) for the 1990s and beyond. The project was the most advanced of its kind, involving an exceptionally manouverable, dynamically unstable, fly-by-wire design, so radical that the new technology was proved in the Experimental Aircraft Programme, EAP. EAP flew for the first time in 1986, one of three maiden flights made by BAe aircraft that year. The others were the Hawk 200, and the ATP which made its first take-off to the very day and hour that had been programmed three years before!

EFA development progressed well and, bearing its new name of Eurofighter 2000, the first prototype flew in March 1994. The development and test programme continues, and Eurofighter 2000 should be in operational service as the new century begins.

The Harrier II Plus in the markings of the US Marine Corps. Ordered by the Spanish and Italian navies, Harrier II Plus will be produced in partnership with McDonnell Douglas in the largest Anglo-US aircraft project ever seen. **(BAe)**

Eight European nations are collaborating in the European Future Large Aircraft Group, formed in 1991. The Royal Air Force has already announced its intention to buy the aircraft which will come in tanker and other variants. **(BAe)**

EUROFLAG srl

The largest collaborative project yet to be undertaken, the European Future Large Aircraft Group (EUROFLAG) was formed in June 1991 and involved eight nations as partners. In June 1995, British Aerospace, Alenia of Italy, Aerospatiale of France, CASA of Spain and Daimler-Benz Aerospace of Germany agreed to form an Airbus military aircraft company prior to FLA's pre-development phase in 1996. The project is to design and develop a large military transport aircraft, with tanker and other variants, as a replacement for transport aircraft currently in service with air forces worldwide. The necessary feasibility studies began in 1993, and once the pre-development phase is complete in 1998, the programme will lead to delivery of the first aircraft in 2003, in line with the requirements of the European air forces.

SAAB MILITARY AIRCRAFT

A joint venture agreement was announced at the Paris Air Show in June 1995 between British Aerospace and Saab Military Aircraft. The new partnership enables British Aerospace to add the Gripen, a new generation multi-role fighter aircraft, to its product range, and Saab to benefit from the worldwide marketing and support experience of British Aerospace. The two companies are working jointly on the adaptation of the Gripen for export markets, with the BAe share of the work being carried out at its Brough site in North Humberside. British Aerospace was involved in the Gripen development programme in the 1980s, when the company designed the aircraft's carbon-fibre wing and manufactured a number of prototype wing sets.

INTO THE 21ST CENTURY

Less than an average lifetime separated Blackburn's move to Brough in 1929, and the announcement that Brough would participate in the development of one of the world's most advanced fighter aircraft in 1995. The progress of aviation in that time has been relentless, and the achievements remarkable, as the boundaries of aircraft design and development have been pushed further and further back. The qualities that put Britain at the forefront of aviation at the beginning of the century, manifestly remain at the century's end. Vision, courage, determination, an entrepreneurial spirit and the pursuit of excellence have never been wanting in the men and women whose efforts brought British Aerospace into being. Those same qualities, exercised in collaboration with colleagues of other nations, will continue to serve British Aerospace well into a new century, and will add further laurels to its already Proud Heritage.

The Saab Gripen which BAe has added to its product range in a joint venture agreement with Saab Military Aircraft. **(BAe)**

The following pages show photographs of most, but not all, of the aircraft that were built by companies which eventually became part of British Aerospace. Not all were successful, and many were not put into production for one reason or another. All, however, made their individual contribution to the fund of knowledge and expertise upon which the British aviation industry continues to draw, and to which it adds the lessons of today.

Great help in the compilation of the photographs has been given by Gordon Bartley of British Aerospace, and the staffs of the Imperial War Museum, the Royal Air Force Museum and the Science Museum. Much of the caption research was completed by Christopher Whitehead, and aided by the work of the late Oliver Tapper in his book Roots In The Sky, to which due acknowledgment is made. The help of Robert Gardner of British Aerospace, whose idea it was to produce this book, is also acknowledged.

Credit for each photograph used has been given in good faith by inserting the initials of the source at the end of each relevant caption. If credit belongs to any person or organisation other than the acknowledged source, due apology is given. Those persons and organisations credited are:

The Science Museum's Science and Society Picture Library **(SSPL)**, The Imperial War Museum **(IWM)**,The RAF Museum **(RAFM)**), British Aerospace **(BAe)**, Adrian Meredith Photography **(AMP)**, Gordon Bartley **(GB)**, David Woolston **(DW)**, Rick Brewell **(RB)**, Geoff Lee **(GL)**.

◀ 1909: ROE I TRIPLANE
To Alliot Verdon Roe goes the credit of making the first all-British aeroplane flight. On 23 July 1909 at Lea Marshes, Essex, his Roe I Triplane made three flights of 900 feet at an altitude of between 10 and 20 feet. The aircraft was powered by a 9 hp JAP air-cooled engine. The flights could have been longer but Roe had not learned how to turn the aircraft and was baulked by a line of trees! Wing span 20 feet; speed 25 mph. (SSPL)

▶ 1909: BLACKBURN MONOPLANE
The First Blackburn Monoplane was a wire-and-kingpost, high-wing machine built for strength. Its 35 hp Green engine proved unable to lift the aircraft for more than very short hops. In attempting a turn on 24 May 1910, one wingtip struck the ground and the aircraft was damaged beyond repair. Wing span 24 ft; maximum speed 60 mph. (SSPL)

◀ 1910: BRISTOL BOXKITE
A copy of the Henri Farman, the 'Boxkite' launched the career of the British and Colonial Aeroplane Company. Powered by a variety of engines ranging from 50 to 70 hp, a total of 76 aircraft was produced. They were used for pilot training at home and abroad. Wing span 34 feet 6 inches; maximum speed 40 mph. (GB)

▶ 1910: ROE III TRIPLANE
Powered by an air-cooled 35 hp JAP engine, the two-seat Roe III first flew on 24 June 1910. Roe could now fly figure-of-eights and duly gained his Aero Club Aviator's Certificate - No 18 - on this aeroplane. Subsequent Roe IIIs were powered by 35 hp Green water-cooled engines. Wing span 31 ft. (GB)

▼ 1911: VICKERS' FIRST
Modelled on a design by Robert Esnault Pelterie (REP), Vickers' first aeroplane had the rear, tubular steel, section built in France, and the rest he completed himself under license. The aircraft crashed soon after its first flight, but was the forerunner of seven other Vickers aircraft of part-metal construction. Wingspan 34 feet 6 inches; weight 1200lb. (IWM)

◀ 1911: AVRO TYPE D
In 1911 Roe produced the stable, easy-to-fly Avro Type D. First flown at Brooklands on 1 April, the Type D was later converted into a floatplane and became the first such type to fly from British waters. Seven Type Ds of varying design were built, three of them being used by the Avro Flying School until 1914. Wing span 31 feet; speed 50 mph. (SSPL)

▶ 1912: AVRO TYPE 500
Orders for the Type 500 from the War Office and the Admiralty were sufficient to enable Roe to re-form his firm as a limited company. The two-seat Type 500 was considered by Roe to be his first successful aeroplane, and was used by the Central Flying School at Upavon. One machine was exported to Portugal. Wing span 36 ft; speed 61 mph. (IWM)

◀ 1912: BLACKBURN SINGLE-SEAT MONOPLANE
Powered by a 50 hp Gnome rotary engine, the Single-Seat Monoplane had a successful career before crashing at Wittering in 1914. The First World War interrupted reconstruction and it was not until 1938 that the remains were discovered and taken to Old Warden. Another war again interrupted work but the Monoplane was eventually restored to full airworthy condition and is regularly displayed by the Shuttleworth Trust. Wing span 32 feet 1 inch; maximum speed 60 mph. (AMP)

▶ 1913: AVRO 504K
First flown in September 1913, the Avro 504 was well in advance of contemporary aircraft. In its J and K versions, it was widely used by the Royal Flying Corps and the Royal Naval Air Service. After the two Services combined in 1918 to form the Royal Air Force, the Avro 504 continued in the training role, principally in the K version, until the late 1920s. Wing span 36 feet; speed 95mph. (BAe)

▶ 1913: SOPWITH BAT BOAT

The Bat Boat was a successful amphibian which first appeared at the 1913 Olympia Air Show. Later that same year it demonstrated its capabilities by winning the Mortimer Singer prize of £500 by making twelve alternate landings on land and water within 3 hours 25 minutes. The time limit set was five hours. The aircraft was subsequently purchased by the Admiralty and used for fleet patrol work in the early part of the First World War. When the Bat Boat took part in the Royal Fleet Review of 1914, it was the first time that the Royal Navy had an aircraft as part of its official review line. A larger and more powerful development was called the Bat Boat II. Wing span 41 feet; maximum speed 65 mph. (IWM)

▶ 1913: SOPWITH TABLOID

Powered by an 80 hp Gnome engine, Sopwith's third aircraft, the biplane Tabloid, could carry two people. It was fast and agile and, during tests at Farnborough, produced impressive performance figures of a speed of 92 mph and an initial rate of climb of 1,200 feet per minute. A Tabloid fitted with floats won the 1914 Schneider Trophy. Wing span 25 feet 6 inches; maximum speed 93 mph. (IWM)

◀ 1913: SOPWITH TRACTOR BIPLANE

Powered by an 80 hp Gnome engine, this aeroplane could seat two passengers in an open cockpit ahead of the pilot. Flown by Harry Hawker, it established a new British altitude record of 11,450 feet in May 1913. Biplanes of this type were supplied to both the RFC and the RNAS, although only the latter saw operational service in France. Wing span 40 feet; maximum speed 74 mph. (IWM)

▶ 1914: VICKERS FB5

The FB5 was a two-seat fighter powered by a 100 hp Gnome Monosoupape rotary engine. The first FB5s were issued to training squadrons in the UK in November 1915. After being withdrawn from operational service, FB5s were used as trainers. Wing span 36 feet 6 inches; maximum speed 70 mph. (IWM)

◀ 1915: DH2

The DH2 was the first British 'fighter'. First flown on 1 June 1915, the 100 hp Gnome Monosoupape or 110 hp Le Rhone engine was mounted behind the pilot, thus providing an uninterrupted forward field of fire for the single Lewis gun. The DH2 first entered service with No 24 Squadron and helped counter the hitherto invincible Fokker monoplane. Four hundred were built. Wing span 28 feet 3 inches; maximum speed 93 mph. (IWM)

▶ 1915: ARMSTRONG WHITWORTH FK3A

The FK 3 was an improved version of the BE2c and was built in large numbers. It was easier to construct and was used in a training role until replaced by the Avro Tutor. The FK3's only operational service was with No 47 Squadron in Salonika. Wing span 40 feet; speed 89 mph. (IWM)

▶ 1915: BRISTOL SCOUT
The Scout was a small biplane and was simple to build. Ordered by both the War Office and the Admiralty, the Scout was the forerunner of the single-seat fighter and 374 were built. Operational service included anti-Zeppelin patrols. Wing span 24 feet 7 inches; maximum speed 110 mph. (IWM)

◀ 1916: TWIN BLACKBURN
Designed for use against Zeppelins, against which it was armed with incendiary steel darts, the Twin Blackburn was the first Blackburn design to go into production and was of unconventional design. It featured a double-fuselage construction, each with a 100 hp Gnome Monosoupape rotary engine. Nine aircraft were built, the last having 110 hp Clerget engines. Also known as the Blackburn TB, the Twin Blackburn was underpowered and, during its short operational life, was never used against the German Zeppelins. Wing span 60 feet 6 inches; maximum speed 86 mph. (IWM)

▶ 1915: SOPWITH BABY
The Baby was widely used by the RNAS for anti-submarine and reconnaissance work. A twin-float seaplane with a third float under the tail, the Baby was a development of the Tabloid and was powered by a 100 hp Gnome rotary or a 110 hp or 130 hp Clerget rotary engine. A Baby fitted with wheels was the first aircraft to take off from the deck of a warship. Wing span 25 feet 8 inches; maximum speed 98 mph. (GB)

◀ 1915: SOPWITH 'ONE-AND-A-HALF STRUTTER'
Named because of the unusual arrangement of the centre-section struts, this aircraft had a variable incidence tailplane and was also fitted with airbrakes. Powered by a 100 hp tractor Clerget engine, it was built in quantity and used as a two-seat fighter reconnaissance aircraft, a single-seat bomber and single-seat fighter by the RFC, RNAS, Belgians, Russians and Italians. It was the first British machine to have a gun firing through the propeller arc. Wing span 33 feet 6 inches; maximum speed 105 mph. (IWM)

▲ 1916: SOPWITH PUP
The Pup reached the Western Front towards the end of 1916 and, despite being powered by an engine of only 80-100 hp, managed to hold its own in dogfighting, mainly due to its ability to maintain height. The Pup was flown by both RFC and RNAS squadrons until the end of 1917. Thereafter, they were used for training. Wing span 26 feet 6 inches; maximum speed 112 mph. (IWM)

▲ 1916: SOPWITH TRIPLANE
The Triplane was a single-seat fighter powered by a 130 hp Clerget 9Z rotary engine and was produced in large numbers for the RNAS. It predated the Fokker Triplane by some months, appearing at the Front early in 1917. It was fully aerobatic but, most importantly, it could outclimb any German fighter. Wing span 26 feet 6 inches; maximum speed 113 mph at 6,500 feet. (IWM)

▲ 1916: SUPERMARINE NIGHT HAWK
Designed as a slow-flying defensive aircraft for use against Zeppelin airships, the Nighthawk was a large quadruplane powered by two 100 hp Anzani radial engines. It was armed with a non-recoil Davis gun, which fired a 1.5lb shell, and two Lewis guns. It also carried a searchlight in the nose. Two were ordered, but only one was built. Wing span 60 feet; maximum speed 75 mph. (IWM)

▶ 1916: ARMSTRONG WHITWORTH FK8

The FK8 was a considerable advance on the FK3 and served in numbers as a reconnaissance and bombing machine. No 35 Squadron was the first to receive the type and took it to France in January 1917. Two Victoria Crosses were won by FK8 pilots in battles against superior odds, a tribute also to the rugged construction of the aeroplane. At the end of the First World War, most FK8s were scrapped, only eight finding their way on to the civilian register. It was an FK8 that flew the first regular airmail service, operated by a small company which was to become the Queensland and Northern Territories Air Service, today well-known as QANTAS. Wing span 43 feet 6 inches; speed 95 mph. (IWM)

◀ 1916: VICKERS FB14

The FB14 was a general-purpose, single-engined tractor biplane powered by a 160 hp Beardmore engine. The aircraft proved underpowered and few of the FB14s produced ever received suitable engines although several alternative installations were tried. In all, 94 airframes were built, most of which were stored. Wing span 39 feet 6 inches; maximum speed 100 mph. (IWM)

▶ 1916 BRISTOL M1c

Despite its excellent performance, the Bristol M1C was unable to overcome official preferences for biplanes and was only built in small numbers. Most were sent to the Middle East because the landing speed of nearly 50 mph was deemed too high for small airfields on the Western Front. Wing span 30 feet 9 inches; maximum speed 130 mph. (IWM)

◀ **1916: DH4**
One of the outstanding aircraft of the First World War, the DH4 day bomber first flew in August 1916. It entered service with No 55 Squadron on 6 March 1917 and was first used operationally one month later. Powered by a number of different engines, the first DH4s were fitted with the BHP engine, the forerunner of the Armstrong Siddeley Puma. With the Rolls-Royce Eagle engine, the DH4 could outstrip the fighters of its day. In all, 1,449 DH4s were built in Britain and a further 4,846 in America for use by the US Army. DH4s were widely used after the war in a wide variety of roles, both military and civilian. Wing span 42 feet 5 inches; maximum speed 143 mph. (IWM)

▶ **1916: ARMSTRONG WHITWORTH FK10**
The FK10 was a small quadruplane designed as a two-seat fighter. Inspiration for the FK10 came from the success of the Fokker and Sopwith Triplanes, but the FK10 had a disappointing performance and few were built. Wing span 28 feet 3 inches; speed 95mph. (IWM)

◀ **1916: DH5**
The introduction of a suitable interrupter gear for the guns enabled the DH5 to combine the superior performance of the tractor biplane with the ability of the pusher to fire forwards. The DH5 was distinguished by the upper wing being staggered 27 inches back from the lower in order to improve the pilots view. Initially employed as a fighter, the DH5 was subsequently used for ground attack duties. 550 were built. Wing span 25 feet 8 inches; maximum speed 109 mph. (IWM)

▶ 1916: BRISTOL FIGHTER F2B

The "Brisfit" was one of the outstanding successes of the First World War, once pilots and gunners had learned how to fight as a team. Production ceased in 1926, after 5,308 F2Bs had been built. F2Bs went out of RAF service in 1932, but those in the RNZAF were not scrapped until 1938. Wing span 39 feet 3 inches; maximum speed 125 mph. (IWM)

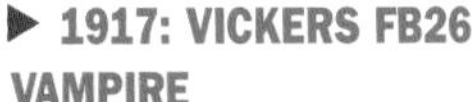

▶ 1917: VICKERS FB26 VAMPIRE

The Vampire was a pusher biplane powered by a 200 hp Hispano Suiza engine. Intended for night-fighting, the prototype FB26 was lost in a crash. Three modified aircraft were built, one of which was armoured for ground attack and powered by a 230 hp Bentley rotary engine. Wing span 31 feet 6 inches; maximum speed 121 mph. (IWM)

◀ 1917: VICKERS FB16D

Powered by a 200 hp Hispano Suiza engine, the FB16D was a single-seat biplane fighter armed with two Lewis guns. Major James McCudden VC DSO MC MM (standing by the FB16D in this photo) flew the aircraft and was delighted by it. It never went into production, however, as large orders had been placed for the SE5a. Wing span 25 feet; maximum speed 135 mph. (IWM)

▶ 1917: SOPWITH DOLPHIN
The first squadron to fly the Dolphin was No 19, in November 1917. Powered by a 200 hp Hispano Suiza engine, it was armed with two fixed Vickers machine guns and one or two semi-free Lewis guns. Visibility was good, and the Dolphin proved to be an excellent fighter, particularly at higher altitudes. Wing span 32 feet 6 inches; maximum speed 128 mph. (IWM)

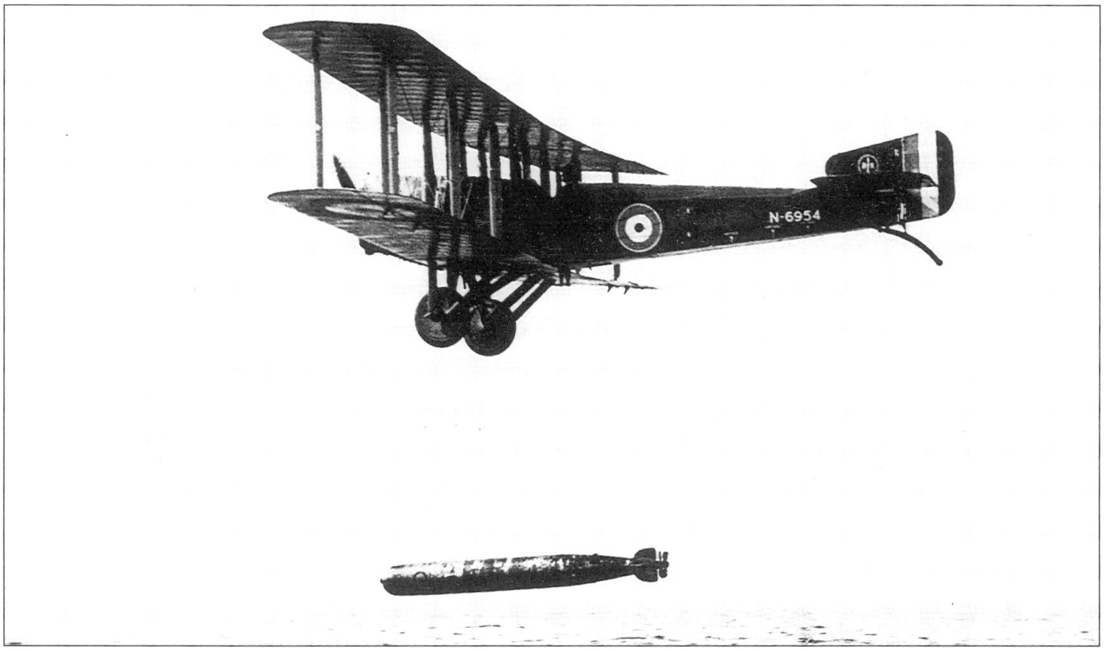

◀ 1917: SOPWITH CUCKOO
A land-based torpedo-bomber, the single-seat Cuckoo was powered by a 220 hp Sunbeam Arab engine. The first aircraft were delivered in June 1918 and thus saw little active service. Some 230 were built and the type was subsequently flown from aircraft carriers. Wing span 45 feet 9 inches; maximum speed 100 mph. (IWM)

▶ 1917: BRISTOL MR1 BIPLANE
Two MR1s were built, the first with an all-metal fuselage and wooden wings, and the second of metal throughout. The design was an insurance against shortage of wood for aeroplane construction. The second aircraft was powered by a 180 hp Wolseley Viper engine. Wing span 42 feet 2 inches; maximum speed 110 mph. (IWM)

▲ 1916: SOPWITH CAMEL
The Camel destroyed more enemy aircraft than any other single type. Used in substantial numbers by both the RFC and the RNAS, the Camel entered service on the Western Front in June 1917. A deadly instrument in the hands of a skilled pilot, it could prove the undoing of the unwary. In October 1918, the RAF had 38 squadrons of Camels. By December 1919, 36 of these had been disbanded. Wing span 28 feet; maximum speed 119 mph. (BAe)

▶ 1917: SOPWITH SNIPE
The last of the small rotary-engined fighters, only three squadrons of Snipes were in time to see action before the end of the war. Developed from the Camel, it remained in front-line RAF service until 1926, some 1,100 being built. In October 1918, Major W G Barker won the VC in a Snipe in combat against 15 Fokker D VIIs. Wing span 30 feet 1 inch; maximum speed 121 mph. (IWM)

◀ 1917: VICKERS VIMY
Designed as a bomber, the Vimy achieved fame by being the first aircraft to cross the Atlantic non-stop in 1919. Too late to see operational service in the war, orders for over 1,000 were drastically reduced to about 100. The Vimy had a range of 1,880 miles and was used on a number of long-distance flights including one from England to Australia. Wing span 68 feet; maximum speed (Mk II) 103 mph. (IWM)

▶ 1917: DH9A
The DH9A was designed by Westland Aircraft who were already experienced in building the DH4 and DH9. The 9A was a DH9 modified to take the American 400 hp Liberty engine. It entered service with No 110 Squadron in August 1918 and remained the RAF's standard day bomber until 1931. Wing span 45 feet 11 inches. (IWM)

◀ 1917: BLACKBURN KANGAROO
Powered by two Rolls-Royce Falcon engines, the Kangaroo could carry four 230 lb bombs or one 520 lb bomb in a special internal compartment, plus four smaller bombs externally. Twenty Kangaroos were built and they flew many convoy protection patrols, during which they attacked several German submarines. On one such sortie a submarine was damaged badly enough for it later to be sunk by depth charges. Surviving Kangaroos went on to the civil register after the war where they were used mainly to give joy-rides, but two Kangaroos went on to compete in the 1922 King's Cup Race. Another entered the race to Australia in 1919 but was forced to retire in Crete. Wing span 74 feet 10 inches; maximum speed 98 mph at 6,000 feet. (IWM)

◀ 1918: SUPERMARINE BABY

A single-seat biplane flying boat, the Baby was designed as a fighter. Powered by a pusher 200 hp Hispano Suiza engine, the Baby featured folding wings and ailerons on all four wings. Only one Baby was built but the design was later successfully developed as a racer. Wing span 30 feet 6 inches; maximum speed 117 mph. (IWM)

▶ 1918: ENGLISH ELECTRIC PHOENIX P5 CORK

Powered by two 360 hp Rolls-Royce Eagle engines, this biplane flying boat first flew on 4 August 1918. The hull was of monocoque construction with two layers of mahogany planking, the outer planks laid longitudinally and the inner diagonally. A second aircraft was subsequently re-engined with 450 hp Napier Lions and a third had Rolls-Royce Condors. Wing span 85 feet 6 inches; maximum speed 105 mph. (IWM)

◀ 1918: BRISTOL BRAEMAR

Designed as a long-range bomber, the first Braemar had four 230 hp Siddeley Puma engines. The type was too late for the war but a second was built, with 400 hp Liberty engines, and converted to produce a single example of the Pullman 14-seat airliner. It was not put into production. Wing span 81 feet 8 inches; maximum speed 125 mph. (IWM)

▶ 1918: DH10A
The DH10A bomber was designed in response to successful German bombing attacks. With two Rolls-Royce Eagle or two Liberty engines, only eight had been delivered to the RAF by the end of the war and the type was subsequently mainly used as a mailplane. Wing span 65 feet 6 inches; maximum speed 129 mph. (IWM)

◀ 1919: DH16
The DH16 was a DH9A with a wider rear fuselage and a glazed cabin able to seat four passengers. Nine DH16s were built, the first six powered by the 320 hp Rolls-Royce Eagle VIII and the last three by the 450 hp Napier Lion. Most were used by Aircraft Transport and Travel Ltd and one inaugurated KLM's London-Amsterdam service in 1920. Wing span 46 feet 6 inches; maximum speed 136 mph. (IWM)

▶ 1919: SOPWITH ATLANTIC
The Atlantic was built to win the £10,000 Daily Mail prize for the first non-stop Transatlantic flight. Its undercarriage was jettisonable to save weight, and an upturned lifeboat formed the rear fuselage decking. It was powered by a 375hp Rolls-Royce Eagle. Wing span 46 feet 4 inches; maximum speed 118 mph. (IWM)

▶ 1919: BRISTOL BADGER F2C

A two-seat fighter-reconnaissance biplane designed for rapid production, the Badger was powered by a 400 hp Cosmos Jupiter radial engine. Like many other promising designs the Badger was not put into production. However, the Jupiter engine was developed by Bristol and, built under license in many countries, became extremely successful. Wing span 23 feet 8 inches; maximum speed 142 mph. (IWM)

◀ 1919: AVRO 534 BABY

Originally named the Popular, the Avro 534 Baby was a return to the light aeroplane able to operate from any reasonably large field. Although several variants were built, and the type scored some competition successes, it was not produced in large numbers. An Avro Aldershot appears with the Baby in this photo. Wing span 25 feet; speed 78 mph. (IWM)

▶ 1919: SOPWITH GNU

One of the first aircraft designed for civil use, the Gnu could seat two passengers in the rear cockpit under a glazed roof. The glazing was unpopular and most of the 12 Gnus built had an open passenger cockpit. Powered by a Le Rhone engine, the Gnu also sold abroad. Wing span 38 feet 1 inch; maximum speed 93 mph. (IWM)

◀ 1919: VICKERS VIKING MK III
A small amphibian, the first Viking flew at Brooklands in late 1919. The Mk III had a 450 hp Napier Lion, and won the 1920 Air Ministry contest for amphibians. It was subsequently used to test the feasibility of a service between the River Thames in London and the River Seine in Paris, and then went aboard HMS Argus for trials. Wing span 46 feet; maximum speed 110 mph. (IWM)

▶ 1920: DH18
This was De Havilland's first aircraft designed for civil use and the first to be produced at the Stag Lane premises. Powered by a 450 hp Napier Lion, the DH18 could carry eight passengers in an enclosed cabin or 2,200 lb of freight. The first of six aircraft was operated by Aircraft Transport and Travel Ltd, and the other five by Instone Airways. Wing span 51 feet 3 inches; maximum speed 128 mph. (AMP)

◀ 1921: AVRO 555 BISON
Designed as a deck-landing reconnaissance and gunnery spotting biplane, the Bison equipped No 3 Squadron from 1922-23. It was used for coastal duties and undertook deck trials on board HMS Furious and HMS Eagle. The Bison was powered by a 480hp Napier Lion engine, and 53 were built. Wing span 46 feet; speed 108mph. (IWM)

◀ 1921: GLOSTER MARS
Planned and constructed in less than four weeks, the Mars was built as a racer to attract publicity for the Gloucestershire Aircraft Company. This it did by winning the 1921 Aerial Derby at a speed of 163.3 mph. The aircraft repeated its success in 1922, winning at 177.85 mph, and again in 1923 when as the Gloster 1 it achieved 192.4 mph. Wing span 22 feet; maximum speed 202 mph. (IWM)

▶ 1922: AVRO 549 ALDERSHOT
Designed as a long-range bomber, the Aldershot was the first Avro type to have a metal fuselage. Two pilots sat side-by-side with the gunner behind while the bomb-aimer and radio operator were accommodated in the spacious cabin below. Fifteen Aldershot Mk IIIs saw service with No 99 Squadron from 1924-26. Wing span 68 feet; speed 110 mph. (IWM)

◀ 1922: DH34
Like its predecessor the DH18, the DH34 was powered by a 450 hp Napier Lion and could carry 10 passengers. It was used by Daimler Hire Ltd, Instone Airways and, later, Imperial Airways on routes from London to Paris, Brussels, Cologne, Amsterdam and Berlin. DH34s were withdrawn from service in 1926. Wing span 51 feet 4 inches; maximum speed 128 mph. (AMP)

▶ 1922: VICKERS VULCAN

The Vulcan was an eight-passenger, single-engined biplane. Three were operated by Instone on their London to Paris or Brussels routes. A further six were built, the last two having the 360 hp Rolls-Royce Eagle engine replaced by a 450 hp Napier Lion. These were used by Imperial Airways. Wing span 49 feet; maximum speed (Lion engine) 112 mph. (AMP)

◀ 1922: VICKERS VIRGINIA

The Virginia was a long-range bomber developed from the Vimy. Entering RAF service in 1924 the Virginia remained a front-line bomber until 1937. A number of marks were built, the most prolific being the Mk X, which was powered by two 580 hp Napier Lions. Virginias were used for various trials including in-flight refuelling. Wing span 87 feet 8 inches; maximum speed (Mk X) 108 mph at 5,000 feet. (IWM)

▶ 1922: VICKERS VICTORIA

This was the transport version of the Virginia bomber and could carry 22 troops. Ninety-four were built and Victorias assisted the evacuation of Kabul during the riots of 1928/29. Powerplants were two 570 hp Napier Lions although the Mk VI had two 622 hp Bristol Pegasus radials. They were replaced by Vickers Valentias in 1936. Wing span 87 feet 4 inches; maximum speed (Mk V) 110 mph. (IWM)

◀ 1922: BLACKBURN BLACKBURN I
The Blackburn was a carrier-borne biplane, designed principally for reconnaissance and gunnery control. The Mk I, of which 33 were built, was powered by a 450 hp Napier Lion and the Mk II, of which there were 30, by a 465 hp Napier Lion. The Blackburn entered service in 1923. Wing span 45 feet 6 inches; maximum speed (Mk I) 122 mph at 3,000 feet. (IWM)

▶ 1922: ARMSTRONG WHITWORTH SISKIN
The Siskin was the RAF's first all-metal aircraft and despite its rather angular appearance formed the main equipment of the RAF's fighter squadrons from 1924 to the early 1930s. Powered by an Armstrong Siddeley Jaguar radial engine, the Siskin was easy to fly compared with earlier fighters. Wing span 33 feet 2 inches; speed 153 mph. (IWM)

▶ 1923: VICKERS VIXEN
The Vixen was a biplane two-seat fighter-bomber which attracted orders from overseas. A number of variants were built and given different names. Chile bought 18 Vixen V aircraft, and Portugal 14 Valparaisos. The RAF bought six Ventures for coastal reconnaissance duties but they were mainly used for experimental flying. Wing span 40 feet; speed 137 mph at 10,000 feet. (IWM)

◀ **1923: SUPERMARINE SEA EAGLE**
The Sea Eagle carried six passengers in a cabin in the front of the fuselage, with the pilot and his mechanic in an open cockpit above them. Three aircraft were built for the Channel Islands service to Guernsey from Southampton. Powerplant was one 360 hp Rolls-Royce Eagle. Wing span 46 feet; speed 93 mph. (AMP)

◀ **1923: ENGLISH ELECTRIC WREN**
A single-seat tractor monoplane, the Wren was essentially a glider powered by a 398cc ABC motorcycle engine. The prototype cost £600 and took only two months to build, performing well in trials. Four Wrens were built one of which was restored to flying condition in 1956. Wing span 37 feet 1 inch; speed 49 mph. (BAe)

▶ **1923: HAWKER WOODCOCK**
The Woodcock was the first Hawker aircraft to be produced in quantity, and equipped Nos 3 and 17 Squadrons as a night fighter. Powered by a 380hp Bristol Jupiter radial engine, the Woodcock entered service in 1925 and was replaced by the Gloster Gamecock in 1928. It was armed with two Vickers machine guns mounted on each side of the fuselage. Wing span 32 feet 6 inches; speed 143 mph. (IWM)

▶ 1923: GLOSTER GREBE
The Grebe was Gloster's first aircraft to be produced in quantity for the RAF and was agreed to be far in advance of any other fighter previously developed. The production model Grebe II was powered by a 400 hp Armstrong Siddeley Jaguar engine and entered service with No 111 Squadron in October 1923. A total of 130 Grebes was built plus a small number of two-seaters - the first time the RAF had had a high performance dual-control trainer. Wing span 29 feet; speed 162 mph. (IWM)

▶ 1923: BRISTOL BULLFINCH
Designed as an all-metal single-seat monoplane fighter, which could be readily converted into a two-seat biplane reconnaissance aircraft, the Bullfinch was powered by a 425 hp Jupiter engine. Two single-seaters were produced and had a good performance but the one biplane built was too heavy. Wing span 38 feet 5 inches; maximum speed (single-seater) 135 mph. (RAFM)

▶ 1924: ENGLISH ELECTRIC M3 AYR
Designed as a single-engined Fleet gunnery and spotting aircraft, the Ayr flying boat had a crew of four. A biplane, the lower wings had exaggerated dihedral and acted as sponsors rendering wingtip floats unneccessary. Powered by a 450 hp Napier Lion, two aircraft were started, but the second was never completed due to the inability of the first to take off. Wing span 46 feet; speed 127 mph. (BAe)

▲ 1924: BLACKBURN CUBAROO
Designed as a long-range coastal defence aircraft able to carry a 21-inch torpedo, the Cubaroo was one of the largest aircraft of its day. Despite this it was powered by just one 1,000 hp Napier Cub 16-cylinder engine. In 1925, the Air Ministry changed its policy towards large single-engined bombers and moved towards twin-engined machines. As a result, only two Cubaroos were built. Span 88 feet; speed 115 mph. (IWM)

▼ 1924: AVRO 504N
The 504N was a radial-engined version of the 504K which was developed after the end of the First World War. It served as a trainer with the RAF until 1933 although, in 1940, seven civil machines were commandeeredfor experimental duties. The 504N was also exported to the air forces of nine countries. Ex-RAF machines were widely used for civilian purposes in the UK. Wing span 36 feet; speed 100mph. (IWM)

▶ 1924: ENGLISH ELECTRIC P5 KINGSTON

Based on the Phoenix P5 Cork, the Kingston was a coastal patrol flying boat. In May 1924 a prototype was ordered by the Air Ministry, but was badly damaged when it struck flotsam during its first take-off run and had to be beached. A number of further aircraft were built for experimental work. Wing span 85 feet 6 inches; maximum speed 105 mph. (IWM)

◀ 1924: SUPERMARINE SWAN

The Swan was the first twin-engined amphibian in the world. Powered by two 360 hp Rolls-Royce Eagle engines, it was designed to carry 12 passengers. Re-engined with two 450 Napier Lions, the Swan was subsequently loaned to Imperial Airways with whom it operated as a ten-seat passenger arcraft on the Channel Islands route. Span 68 feet 8 inches; speed 109 mph. (IWM)

▶ 1925: HAWKER HORSLEY

This was the last all-wooden aircraft built by Hawker. Later marks were of metal and wood construction, and the type was operated by the RAF both as a bomber and as a torpedo-bomber. Powered by a 665 hp Rolls-Royce Condor engine, the Horsley was also used by the Greek Naval Air Service. A number of Horsleys were used as engine test beds. Wing span 56 feet 6 inches; maximum speed 125 mph. (IWM)

◀ 1925: GLOSTER GAMECOCK
Despite having an excellent performance, the Grebe suffered from the unreliability of its Jaguar engine. The answer was the new Bristol Jupiter which was fitted to a development of the Grebe named the Gamecock. First deliveries were to No 23 Squadron in May 1926 and 88 were produced for the RAF as its last wooden fighter. Gamecocks were also used by the Finnish Air Force. Span 29 feet 10 inches; speed 145 mph. (RAFM)

▶ 1925: ARMSTRONG WHITWORTH ATLAS
Designed as a two-seat army co-operation biplane, the Atlas followed the Siskin, a total of 478 being built. Like the Siskin, it was of all-metal construction and was powered by an Armstrong Siddeley Jaguar engine. Fitted to the landing gear it had a retractable hook which could be lowered for picking up message bags from the ground. In later years the Atlas served as an advanced trainer before being phased out in 1935. The Royal Canadian Air Force, however, was still operating the Atlas in 1939. Wing span 39 feet 7 inches; speed 142 mph. (IWM)

◀ 1925: DH60 MOTH
The Moth was a sturdy two-seat biplane designed for club and private owner use. First flown on 22 February 1925, it was an instant success and equipped five Air Ministry-subsidised flying clubs. A large number of Moths were sold all over the world and the aircraft was produced in a number of variants powered by Cirrus engines of between 60 and 105 hp. Wing span 29 feet; speed 95 mph. (BAe)

▶ 1925: SUPERMARINE SOUTHAMPTON
Two 500 hp Napier Lion engines powered this long-range flying boat which was ordered by the RAF straight off the drawing board. The Mk I had a wooden hull but the Mk II's metal hull was some 900 lb lighter. Southamptons made many long-distance flights and equipped five RAF squadrons. Wing span 75 feet; maximum speed 95 mph. (IWM)

◀ 1925: GLOSTER III
One of the two Gloster III racing seaplanes came second in the 1925 Schneider Trophy contest. It was powered by a 700 hp Napier Lion twelve-cylinder engine which enabled the aircraft to achieve an average speed of 199.16 mph in the race. Both Gloster IIIs were subsequently used by the RAF's High Speed Flight for development and training. Wing span 20 feet; maximum speed 225mph. (SSPL)

▶ 1925: VICKERS VESPA
The Vespa was designed as an army co-operation aircraft with a view to replacing the Bristol Fighter in that role. Six Vespas were delivered to Bolivia where its high-altitude performance was put to good use. A Vespa fitted with a Bristol Pegasus 'S' supercharged engine took the World's Height Record by reaching 43,976 feet. Span 50 feet; speed 149 mph. (IWM)

◀ **1926: ARMSTRONG WHITWORTH ARGOSY**
Powered by three Armstrong Siddeley Jaguar engines, the Argosy was operated by Imperial Airways on its London-Paris route. Later aircraft were used on routes in the Middle East. The Argosy could carry 20 passengers, each of whom had an opening window to hand! Wing span (Mk II) 90 feet 4 inches; speed 110 mph. (AMP)

▲ **1926: BLACKBURN IRIS**
Powered by three 650 hp Rolls-Royce Condor III engines, the Iris I was a long-range reconnaissance flying boat. Five marks in all were designed but only 10 aircraft were produced. Nevertheless, the Iris made some notable long-distance flights while in RAF service. Wing span 95 feet 6 inches; maximum speed (Mk I) 115 mph at sea level. (IWM)

◀ **1926: BLACKBURN RIPON**
The main production versions of the Ripon were the Mks II, IIA and IIc, 92 of these being built. A two-seat torpedo reconnaissance aircraft, the Ripon replaced the Dart in Fleet Air Arm service in 1929. It was gradually replaced from January 1934 by the Baffin. A further 25 Ripons were built under licence in Finland, some being completed as floatplanes. Wing span 44 feet 10 inches; speed 132 mph. (RAFM)

▶ **1926: AVRO 594 AVIAN**
This pretty little biplane was produced with a number of different engines and saw widespread service with private owners and flying clubs in the UK and overseas. Avians broke the UK/Australia record in 1928 and again in 1930. Wing span 28 feet; maximum speed (Mk IV) 87 mph. (IWM)

◀ **1927: GLOSTER GORING**
The Goring was built as a private venture and was a potential replacement for the Hawker Horsley. It was a two-seat day bomber/torpedo aircraft powered by a 470 hp Jupiter VII engine. This proved unreliable and the sole Goring built subsequently appeared with a number of different engines. The Goring was also fitted with floats and tested as a seaplane. Span 42 feet; speed 136 mph. (RAFM)

◀ **1927: BRISTOL BULLDOG**

The Bulldog was built as a fast light fighter able to catch the new generation of fast bombers. It was selected for RAF service and over 400 were built. Bulldogs were exported to eight other countries and it was undoubtedly one of the finest aerobatic biplanes ever made. It was powered by a 440 hp Bristol Jupiter radial engine. Wing span 33 feet 10 inches; maximum speed 178 mph. (RAFM)

◀ **1927: SUPERMARINE S5**

Developed from the S4 for the 1927 Schneider Trophy race, the S5 had an all-metal fuselage and wooden wings. Two were built and they finished first and second in the contest, the winning speed being 281.65 mph. They were each powered by a Napier Lion engine which developed 900 hp at 3,300 rpm. Wing span 26 feet 9 inches; speed 319.57 mph. (BAe)

▶ **1928: VICKERS VILDEBEEST**

Developed from the Vixen, the Vildebeest was a large biplane torpedo-bomber powered by a 660 hp Bristol Pegasus radial. 168 were built for the RAF and the aircraft was still in service in 1942. The Royal New Zealand Air Force bought 27, and 25 were built under licence for the Spanish Navy. The Vincent was one of the variants. Span 49 feet; speed 143 mph. (IWM)

▶ 1928: SUPERMARINE SOLENT

The Solent was a three-engined version of the Southampton and was originally built as a torpedo-carrier for the Danish Navy. The order was later cancelled, so the aircraft was finished as a luxurious 12-seater for the Hon. A E Guinness. Powered by three 430 hp Armstrong Siddeley Jaguars, Solents were used for flights between Southampton and Ireland. Wing span 75 feet; maximum speed 130 mph. (IWM)

◀ 1928: HAWKER TOMTIT

Designed to replace the Avro 504N, the Tomtit was an all-metal elementary trainer powered by a 150 hp Armstrong Siddeley Mongoose engine. Despite its excellent flying characteristics, only a limited number of Tomtits were built for RAF and civilian use. Wing span 25 feet 7 inches; maximum speed 124 mph. (GB)

▶ 1929: VICKERS TYPE 143

The Type 141 Scout was entered in the single-seat fighter competition which was won by the Bulldog. The Type 143 was a development of this aircraft, and six were ordered by Bolivia in 1929. Powered by the 450 hp Bristol Jupiter radial engine, the Type 143 was one of a number of experimental Vickers fighters. Wing span 34 feet; maximum speed 150 mph. (IWM)

◀ **1929: SUPERMARINE SEAMEW**
A small twin-engined amphibian designed for shipboard use, the Seamew was powered by two Armstrong Siddeley Lynx radials and had a crew of three. The Seamew was the first Supermarine design to use metal in its construction. Two aircraft were built but were scrapped after a relatively short life. Wing span 46 feet; maximum speed 95 mph. (IWM)

▶ **1929: SUPERMARINE S6**
The S6 was a direct development of the 1927 Schneider Trophy-winning S5, and was powered by the new Rolls-Royce 'R' engine which produced 1,900 hp at 2,900 rpm. One of the two S6s entered won the 1929 race at a speed of 328.63 mph. The same aircraft also set a new world airspeed record of 357.7 mph. Span 30 feet; speed 357.7 mph. (BAe)

◀ **1928: HAWKER HART**
Designed as a bomber, the Hart was also used as an advanced trainer. From it were developed the Audax army co-operation aircraft, the Hind bomber, the Demon two-seat fighter, the Hardy general-purpose aircraft and the Osprey for carrier operations. The Hart entered RAF service with No 33 Squadron in January 1930, and one was still being used in 1943. Wing span 37 feet 3 inches; speed 184 mph. (GB)

◀ 1929: AVRO TUTOR
A worthy successor to the 504 it was designed to replace, the Tutor became the standard trainer for the RAF and more than a dozen overseas air forces. It was powered by the reliable Armstrong Siddeley Lynx engine, and remained in RAF service until 1938. Wing span 34 feet; speed 128 mph. (IWM)

▶ 1929: DH80 PUSS MOTH
The growth of private flying led to the Puss Moth, which was faster than the DH60 Moth and more comfortable with its enclosed cabin. It was in production for three years, and nearly 50% of those built were exported. The type was flown by the Mollisons and by Amy Johnson on many long-distance flights. Brought into RAF service in 1939 for communications work, a few returned to the civil register in 1945. Span 36 feet 9 inches; speed 128 mph. (IWM)

◀ 1930: BLACKBURN SYDNEY
The Sydney was Britain's first large monoplane flying-boat and was powered by three 525 hp Rolls-Royce F.XII engines. Only one aircraft was built and this was used mainly for research purposes until 1934. Wing span 100 feet; maximum speed 123 mph at 5,000 feet. (IWM)

▶ 1929: SUPERMARINE AIR YACHT

This was Supermarine's first multi-engined monoplane. It originated as a three-engined reconnaissance flying boat for the Air Ministry but was subsequently built as a luxury six-seater private aircraft for the Hon A E Guinness to replace his Solent flying boat. Wing span 92 feet; speed 111mph. (RAFM)

◀ 1931: HAWKER FURY

A single-seat biplane interceptor fighter, the Fury was the RAF's first front-line aircraft able to exceed 200 mph in level flight. Powered by a 525 hp Rolls-Royce Kestrel engine, the Fury entered service with No 43 Squadron in May 1931. The Mk II had the more powerful 640 hp Kestrel engine and served with the RAF until 1939. The Nimrod was a naval version. Wing span 30 feet; maximum speed (Mk II) 223 mph. (RAFM)

▶ 1931: SUPERMARINE S6B

Supermarine was able to build a contender for the 1931 Schneider Trophy due to the generosity of Lady Houston who contributed £100,000. The S6B duly took the Trophy, flying the course at an average speed of 340 mph. This triumph gained the Schneider Trophy outright for Britain and paved the way for the development of the Spitfire. Wing span 30 feet; maximum speed 407.5 mph. (RAFM)

▶ 1931: DH TIGER MOTH
One of the finest training aircraft of all time, the Tiger Moth was developed directly from the DH60 Moth. Powered by an inverted Gypsy Major engine, it became the RAF's standard elementary trainer, the last being replaced by Chipmunks in 1955. Over 8,000 Tiger Moths were built and they have been operated all over the world. Wing span 29 feet 4 inches; speed 90 mph. (GB)

◀ 1932: SUPERMARINE SCAPA
A successor to the Southampton, the Scapa was powered by two 525 hp Rolls-Royce Kestrel engines. An all-metal aircraft with fabric-covered flying surfaces, 12 were ordered by the Air Ministry and they served with the RAF until 1939. Wing span 75 feet; maximum speed 142 mph. (IWM)

▶ 1932: GLOSTER GAUNTLET
The advent of the Fairey Fox bomber capable of 156 mph caused acute embarrassment to fighter squadrons which were unable to catch it. After a lengthy development period, Gloster produced the Gauntlet which entered service with No 19 Squadron in May 1935. Powered by a Bristol Mercury engine the aircraft was capable of well over 200 mph. Span 32 feet 10 inches; maximum speed 230 mph. (IWM)

◀ **1932: BLACKBURN BAFFIN**

Continuing the range of Blackburn torpedo-bomber types, the Baffin was the successor to the Ripon. After trials with alternative engines, the Bristol Pegasus was selected and the type went into service with the Fleet Air Arm aboard HMS Glorious in 1934. The Royal New Zealand Air Force used the Baffin operationally in the Pacific area until 1941. The Baffin was very similar to the Ripon apart from the engine; indeed a number of Ripons were converted into Baffins. Wing span 44 feet 10 inches; maximum speed 125 mph at 6,500 feet. (IWM)

▶ **1933: SUPERMARINE WALRUS**

Built as the Seagull V and ordered by the Royal Australian Navy, this biplane amphibian was powered by a pusher 775 hp Bristol Pegasus radial engine. In RN and RAF service, it was named Walrus. The RAF used it from 1941 onwards for air-sea rescue duties. Over 700 were built. Wing span 45 feet 10 inches; maximum speed 135 mph. (IWM)

◀ **1933: BLACKBURN PERTH**

Four Perths were built and they were the largest biplane flying-boats ever used by the RAF, replacing the Iris Vs of No 209 Squadron. Powered by three 825 hp Rolls-Royce Buzzard engines, the Perth carried a 37mm anti-shipping canon mounted in the bow. Wing span 97 feet; maximum speed 132 mph at sea level. (IWM)

◀ 1933: AIRSPEED AS5 COURIER
The Courier was a fast small-capacity transport and was the first British aircraft to feature a retractable undercarriage. Sixteen were built. One was used in air-to-air refuelling trials and another finished third in the handicap section of the 1934 London-Melbourne race. The AS5A was powered by a 240 hp Lynx IVC radial engine. Wing span 47 feet; speed 154 mph. (IWM)

▶ 1933: BLACKBURN SHARK
Another Blackburn torpedo bomber designed for carrier operations, the Shark was powered by various marks of the Armstrong Siddeley Tiger or Bristol Pegasus engines. The Fleet Air Arm received 239 Sharks many of which were seaplanes. Sharks also served with the Portugese Navy and were built under licence in Canada. Most Sharks had been relegated to target-training by the outbreak of war. Span 46 feet; speed 150 mph. (BAe)

◀ 1934: DH86 AIRLINER
Designed and built in four months to an Australian requirement, the DH86 was a fast ten-seater airliner powered by four Gipsy Six six-cylinder engines. A total of 62 aircraft was built, and operators included Quantas, Imperial Airways, Railway Air Services and Jersey Airways. Wing span 64 feet 6 inches; maximum speed 145 mph. (AMP)

▶ 1934: DH89 DRAGON RAPIDE

The DH89 was a twin-engined, scaled-down version of the DH86, and 728 were built. The Rapide was powered by two 200hp Gipsy Six engines and was widely used for airline and charter work. Rapides saw worldwide service, including RAF use as a trainer and transport where it was known as the Dominie. The DH89 had a single pilot and could carry up to eight passengers. Wing span 48 feet; maximum speed 157 mph. (AMP)

▲ 1934: DH88 COMET

De Havilland's built the Comet to try to ensure that Britain won the 1934 England-Australia race, for which the prize money was £15,000. A small streamlined low-wing monoplane with a retractable undercarriage, the Comet was powered by two DH Gipsy Six engines. Three Comets were entered for the race and finished first and fourth, with one retiring. Two further DH88s were built, one of which was used as a mailplane by the French Government. Wing span 44 feet; maximum speed 237 mph. (RAFM)

▶ 1935: SUPERMARINE STRANRAER

The Stranraer was a larger version of the Scapa flying boat and was powered by two 920 hp Bristol Pegasus engines. With a crew of six the Stranraer could carry two 250lb bombs. First flown in July 1934, it served with the RAF until October 1942 and with the Royal Canadian Air Force until April 1944. Wing span 85 feet; maximum speed 165 mph. (IWM)

◀ 1934: GLOSTER GLADIATOR

The Gladiator was the last and most advanced of the RAF's biplane fighters. Developed from the Gauntlet, the Gladiator was easily distinguishable by its closed cockpit. More than 740 Gladiators were built, and the type retired from RAF front-line service in September 1941. Wing span 32 feet 3 inches; speed 257 mph. (GB)

▶ 1935: AVRO ANSON

Originally designed as a small airliner, the Anson became the RAF's first operational monoplane when it entered service in 1936 as a coastal reconnaissance aircraft. Withdrawn from front-line service in 1942, it was then used for training and as a light transport. Over 11,000 were produced, the last delivery being made to the RAF on 27 May 1952. Span 57 feet 6 inches; maximum speed 171 mph. (BAe)

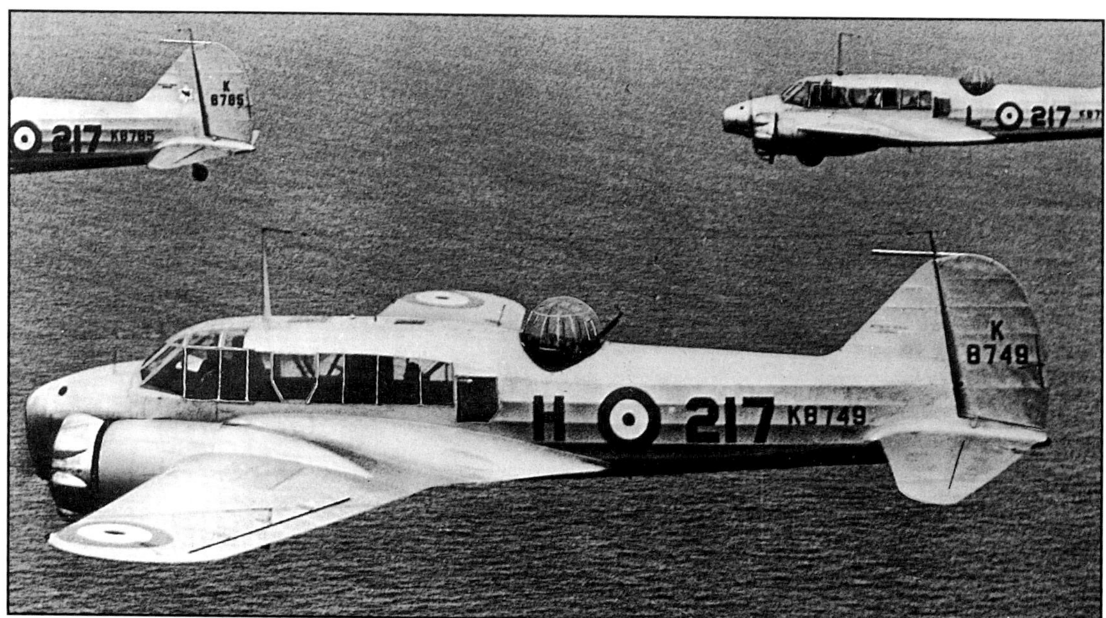

◀ **1936: VICKERS WELLESLEY**
Built using Barnes Wallis' geodetic construction, the Wellesley was a single-engine monoplane bomber, powered by a 925 hp Bristol Pegasus radial. 96 were ordered by the Air Ministry in September 1935 and the first squadron to receive them was No 76. A further seven squadrons were equipped with the Wellesley and the aircraft was used for long-distance research flights. 176 were built, and the Wellesley finally retired in 1943. Wing span 74 feet 7 inches; maximum speed 228 mph. (IWM)

▲ **1935: HAWKER HURRICANE**
Entering service in December 1937 with No 111 Squadron, the Rolls-Royce Merlin-powered Hurricane was the RAF's first monoplane fighter and was armed with eight .303 Browning machine guns. A rugged design, the Hurricane was the RAF's most numerous fighter during the Battle of Britain in 1940. It was subsequently used as a 'tank-buster' and, armed with rockets, as a ground attack aircraft. Hurricanes also operated off small escort carriers and some were catapulted off merchant ships. More than 14,000 were built in England and Canada. Wing span 40 feet; maximum speed (Mk IIc) 336 mph. (IWM)

◀ 1935: BRISTOL BOMBAY
The Type 130 Bombay was able to carry 24 fully-equipped troops or be used as a long-range bomber. Powered by two 1,010 hp Bristol Pegasus engines, it had a range of 2,230 miles and equipped four RAF squadrons, as a transport and night-bomber, mainly in the Middle East. Fifty were built. Wing span 95 feet 9 inches; maximum speed 192 mph. (IWM)

▶ 1936: SUPERMARINE SPITFIRE
Developed from the S-series of Schneider Trophy seaplanes, the Spitfire was a single-seat fighter carrying eight .303 Browning machine guns. Its Rolls-Royce Merlin engine was developed from the Rolls-Royce 'R' racing engine. Between 1936 and 1954, over 20,000 Spitfires in more than 30 variants were produced. Its naval equivalent was the Sea Fire. Span 36 feet 10 inches; speed (MkI) 364 mph. (PM)

◀ 1936: ARMSTRONG WHITWORTH WHITLEY
The Whitley was a heavy bomber ordered off the drawing board. Powered originally by two Armstrong Siddeley engines, later marks were fitted with Rolls-Royce Merlins and could carry 7,375lb of bombs. Whitleys were also used by Coastal Command and as glider-tugs by the Airborne Forces. Span 84 feet; speed 244 mph. (IWM)

▲ **1936: BRISTOL BLENHEIM**

Developed from the Type 142 Britain First, the Blenheim was a twin-engined medium bomber with a crew of three. Powered by two Bristol 840 hp Mercury engines, the Blenheim Mk1 entered RAF service in March 1937. Later versions had a longer nose and the more powerful 920 hp Mercury engines. Blenheims were also used as fighters, in which role they carried a pack of four .303 Browning machine guns under the fuselage. The Blenheim was aslo built in Canada, Finland and Yugoslavia. Wing span 56 feet 4 inches; maximum speed (Mk IV) 295 mph. (GB)

▲ **1937: DH91 ALBATROSS**

Powered by four 525 hp Gipsy Twelve engines, the graceful Albatross was designed as a competitor to the Douglas DC-2 airliner. They carried sufficient fuel to enable them to fly 2,500 miles with a payload of 1,000 lb at 210 mph against a headwind of 40 mph. Two experimental mailplanes were built followed by five production 22-passenger airliners for Imperial Airways. Wing span 104 feet 8 inches; maximum speed 225 mph. (AMP)

▲ 1936: VICKERS WELLINGTON

The Wellington was the mainstay of Bomber Command in the early years of the war pending the arrival of the four-engined bombers. First flown on 15 June 1936, the Wellington's geodetic structure made it immensely strong. A number of variants were built and powered by a pair of Bristol Pegasus, Rolls-Royce Merlin or Bristol Hercules engines. They were used by Bomber, Coastal, Transport and Flying Training Commands, total production amounting to 11,461. Wellingtons were still in front-line service as late as March 1945. Wing span 86 feet 2 inches; maximum speed (Mk IE) 235 mph at 15,500 feet. (BAe)

▶ 1936: PERCIVAL Q6 PETREL

Percival's first twin-engined transport, built to carry five or six passengers, the Petrel was powered by Gipsy Six engines and was offered with an optional retractable undercarriage. A few were sold abroad, including one to the King of Iraq, but the outbreak of war limited the production run to only 15 aircraft. Seven were delivered to the RAF, which also impressed three civil Q6s during the war. The Petrel in the foreground of this photograph is accompanied by Bristol Blenheims behind. Wing span 46 feet; maximum speed 195 mph. (IWM)

▲ 1937: BLACKBURN SKUA

The Skua was a monoplane two-seat fighter/dive bomber, the first such type to enter Fleet Air Arm service. It was also the first British deck-landing aircraft equipped with flaps, a retractable undercarriage and variable-pitch propellor. One hundred and ninety were built and its greatest feat was to sink the German light cruiser Konigsberg by dive bombing. Wing span 46 feet 2 inches; maximum speed 225 mph at 6,500 feet. (IWM)

▲ 1937: AIRSPEED QUEEN WASP

Powered by a 350 hp Armstrong Siddeley Cheetah IX radial engine, the Queen Wasp was a radio-controlled pilotless target. Both land and floatplane prototypes were built but only four or five production aircraft were delivered to the RAF before the need for it had passed. Wing span 31 feet; maximum speed 172 mph at 8,000 feet. (IWM)

▶ **1937: AIRSPEED OXFORD**

The Oxford was a twin-engined advanced trainer and was the first British aircraft to be designed with a logically planned cockpit layout. 8,500 Oxfords were built and were used by the RAF and air forces of the Commonwealth. Most were powered by the Armstrong Siddeley Cheetah engine, but some had 450hp Pratt and Whitney radials. The Oxford was developed from the civil Envoy, and those on the civil register after the war were named Consuls. Wing span 53 feet 4 inches; maximum speed (Mk I) 182 mph at 8,300 feet. (IWM)

▲ **1938: ARMSTRONG WHITWORTH ENSIGN**

A 40-seat four-engined airliner, the Ensign was designed for use by Imperial Airways in Europe and in the East to supplement its flying boat services. The original 850 hp Tiger engines did not develop enough power and Ensigns were re-engined with 950 hp Wright Cyclones. Fourteen Ensigns were built and saw considerable service during the war, two being captured by the Germans and refitted with Daimler-Benz engines. Wing span 123 feet; maximum speed 208 mph at 7,200 feet. (AMP)

◀ 1938: DH95 FLAMINGO

The Flamingo was De Havilland's first stressed-skin all-metal aircraft and was intended for short-haul operations. Able to carry up to 17 passengers, the promising future of the Flamingo was cut short by the Second World War and only 16 were built. They were used by the RAF as VIP transports, by the Royal Navy and by BOAC in the Near East. Wing span 70 feet; speed 239 mph. (IWM)

▶ 1938: SUPERMARINE SEA OTTER

An improved version of the Walrus powered by a 965 hp Bristol Mercury tractor engine, the Sea Otter was the last of a long line of Supermarine biplane amphibians and was also the last biplane to enter RAF service. Sea Otters were used by both the RAF and Fleet Air Arm, 290 being built. After the war several aircraft were transferred to Denmark, the Netherlands and France. Wing span 46 feet; maximum speed 163 mph. (IWM)

◀ 1939: PERCIVAL PROCTOR

The Proctor was a three-seat version of the Vega Gull, and production of all five of its variants amounted to 1,300. They were used by the RAF for communications and training. The Proctor 4 reverted to a four-seater and the Proctor 5 was a postwar civil version. Span 39 feet 6 inches; speed 165 mph. (IWM)

▲ **1938: BRISTOL BEAUFORT**

The Beaufort was derived from the Blenheim and was a torpedo-bomber carrying a crew of four. Powered by two 1,130 hp Bristol Taurus engines, the Beaufort entered RAF service with No 22 Squadron in January 1940. Later marks of Beaufort had two 1,200 hp Pratt & Whitney Twin Wasp engines. Some 1,429 Beauforts were built in England and a further 700 in Australia. Wing span 57 feet 10 inches; maximum speed 265 mph. (IWM)

▶ **1939: VICKERS WARWICK**

The Warwick was designed as a heavier version of the Wellington but problems with the Rolls-Royce Vulture engines delayed production until 1942. By this time, the four-engined bombers were coming into service and there was no need for the Warwick as a bomber. Later, and powered by two Pratt & Whitney Double Wasp or two Bristol Centaurus engines, the Warwick was used for transport, and air-sea rescue duties when it carried a lifeboat which could be dropped to survivors. Span 96 feet 9 inches; speed 224 mph. (IWM)

▶ 1939: BRISTOL BEAUFIGHTER
The Beaufighter was a two-seat fighter which used the wings, tail unit and undercarriage of the Beaufort to speed up production. It was powered by two 1,400 hp Bristol Hercules engines, although one mark had Rolls-Royce Merlins. Early Beaufighters carried four cannon and were used for night-fighting, aided by the first airborne interception radars. Beaufighters saw service in all theatres of war as a strike-fighter armed with rockets and as a torpedo bomber. More than 5,500 Beaufighters were built in England, and a further 364 in Australia. Wing span 57 feet 10 inches; maximum speed 330 mph. (IWM)

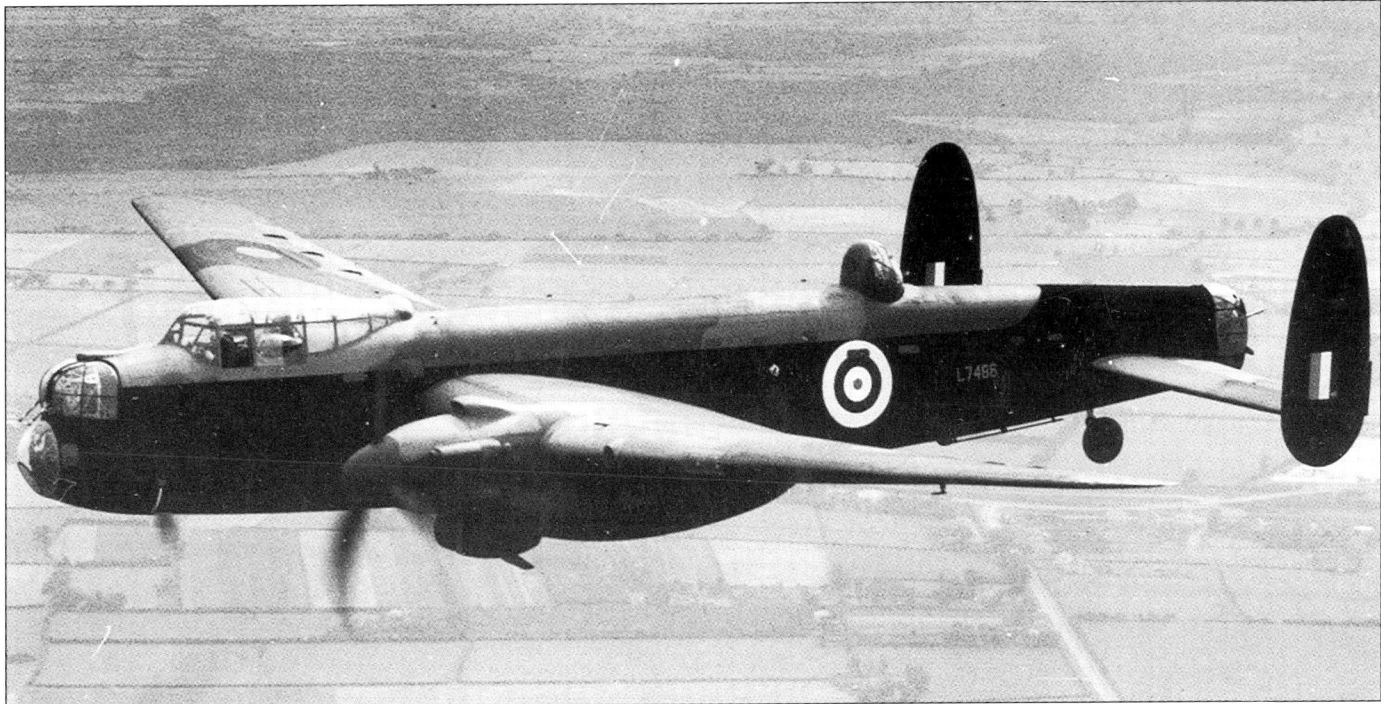

▲ 1939: AVRO MANCHESTER
Designed as a medium bomber, the Manchester entered RAF service in 1940 and equipped 12 squadrons. Its two Rolls-Royce Vulture engines proved unreliable, and Manchesters were withdrawn from service in 1942. However, the design led directly to that of the excellent Lancaster. Wing span 90 feet 1 inch; maximum speed 265 mph. (IWM)

▲ **1940: HAWKER TYPHOON**

Designed as a successor to the Hurricane, the Typhoon was powered by a Napier Sabre engine. Suffering some early structural problems, the Typhoon nevertheless became one of the RAF's most successful ground-attack fighters and carried cannon, bombs and rockets. The Typhoon entered service with No 56 Squadron in September 1941 and more than 3,300 were built. Wing span 41 feet 7 inches; maximum speed 412 mph. (IWM)

▲ **1940: ARMSTRONG WHITWORTH ALBERMARLE**

Designed as a reconnaissance-bomber, the Albemarle was built of wood and steel and was sub-contracted to a wide number of firms as an insurance against a shortage of alloys or the destruction of aircraft factories. Powered by two Bristol Hercules 1,590 hp engines, the Albemarle was used only as a glider tug and airborne forces transport. A total of 608 was built. The aircraft in the photo is unusual as it is in Russian markings. Wing span 77 feet; maximum speed 265 mph at 10,500 feet. (IWM)

◀ 1940: DH 98 MOSQUITO

An outstanding aircraft, the Mosquito was known as the 'Wooden Wonder' because of its all-wood construction. Despite considerable official misgivings, a contract for 50 aircraft was eventually placed and the prototype was built and flown less than 11 months after design work was started. In all, 7,781 were built in many different marks. They were used as light bombers, night fighters, strike fighters and photo-reconnaissance aircraft. The last Mosquito was retired from RAF service in December 1955. Wing span 54 feet 2 inches; maximum speed (PR Mk IX) 408 mph. (BAe)

▲ 1941: AVRO LANCASTER

The outstanding bomber of the Second World War, the Lancaster was developed directly from the Manchester but had four of the well-tried Rolls-Royce Merlin engines. Early Lancasters could carry 7,000 lb of bombs but subsequent developments enabled it to lift the 22,000 lb 'Grand Slam' bomb. Over 7,000 Lancasters were built, including 300 Mk IIs powered by Bristol Hercules radial engines. Wing span 102 feet; speed 287 mph. (IWM)

◀ 1941: GLOSTER E28/39
Britain's first jet-powered aeroplane had a Whittle W.1 engine, and first flew from RAF Cranwell on 15 May 1941. A second aircraft was powered by the 1,200 lb st Rover W.2B but was later lost when the ailerons jammed. In April 1946 the first prototype was put on permanent display in the Science Museum, London. Wing span 29 feet; speed 466 mph. (IWM)

▶ 1941: AIRSPEED HORSA
Designed and flown in less than ten months, the Horsa was a troop and vehicle-carrying glider which took part in the invasions of Sicily and Normandy, the Battle of Arnhem and the Rhine crossing. The Mk I could carry 25 troops while the Mk II had a hinged nose which enabled direct loading of small vehicles and guns. Nearly 3,800 Horsas were built. Span 88 feet; maximum towing speed 150 mph. (IWM)

◀ 1942: HAWKER TEMPEST
Developed from the Typhoon, the Tempest incorporated a thinner wing. The first Tempests were powered by the 2,180 hp Napier Sabre engine, and these aircraft were designated Mk V. Later Tempests had a Bristol Centaurus radial engine and this variant was designated Mk II. Tempests were used in defeating the threat of the V-1 flying bomb. Span 41 feet; speed 442 mph. (IWM)

▲ 1942: BLACKBURN FIREBRAND

Originally conceived as a short-range fleet fighter in 1939, the Firebrand suffered from a succession of changes of policy and finally entered service in 1945 as a torpedo-carrier. Later marks were powered by a 2,520 hp Bristol Centaurus engine which gave the aircraft an impressive performance. Two hundred and twenty Firebrands of all marks were built before the type was withdrawn from service in 1953. Wing span 51 feet 3 inches; maximum speed 342 mph. (IWM)

▲ 1942: AVRO YORK

The York was a transport aircraft with a new square-section fuselage but utilising the wings, tail, undercarriage and engines of the Lancaster. It was not produced in large numbers during the war due to the agreement between Britain and America that the latter would be largely responsible for transport aircraft. Eventually, 256 were built and both civil and RAF versions gave valuable service during the Berlin Air Lift. Wing span 102 feet; maximum speed 298 mph. (AMP)

▶ 1943: DH 100 VAMPIRE
The Vampire first flew on 29 September 1943. Powered by a 2,700 lb st Goblin turbojet, the Vampire was the first aircraft to exceed 500 mph in level flight but was too late for the Second World War. Later, the Vampire was used by the RAF as a day and night fighter and as a trainer, and by the RN as the Sea Vampire fighter. It was also widely exported. Span 40 feet; speed 540 mph. (IWM)

◀ 1943: VICKERS WINDSOR
Designed as a heavy bomber, this aircraft was basically a four-engined Warwick in much the same way as the Lancaster was a four-engined Manchester. It was powered by 1,635 hp Rolls-Royce Merlins and each of the four engine nacelles housed a main undercarriage leg. Too late for wartime service, production orders for the Windsor were cancelled. Span 117 feet 2 inches; speed 317 mph. (IWM)

▶ 1943: GLOSTER METEOR
Powered by two Rolls-Royce Welland turbojets, the Meteor F.1 was the RAF's first jet fighter. It entered service with No 616 Squadron on 12 July 1944 and shot down its first V-1 flying bomb on 4 August. In development and production until 1955, 3,545 of all marks were built. They were used by air forces all over the world. Wing span (F.8) 37 feet 2 inches; speed (F.8) 598 mph. (BAe)

◀ 1944: SUPERMARINE SPITEFUL
The Spiteful was a follow-on design from the Spitfire and featured a new laminar-flow wing. Powered by a 2,375 hp Rolls-Royce Griffon, the Spiteful was too late for the war and only 17 were built, further orders being cancelled in favour of jets. The Spiteful's naval counterpart was the Seafang, but only 11 of these were completed and flown. Wing span 35 feet 6 inches; maximum speed 475 mph at 28,500 feet. (RAFM)

▶ 1944: AVRO LINCOLN
An improved Lancaster, the Lincoln was designed to have a very long range for use in the Pacific theatre. It arrived too late to see operational service there, but did take part in later operations in Malaya and Kenya. It remained the RAF's front-line bomber until replaced by the Canberra, starting in 1955. Wing span 120 feet; speed 295 mph. (RAFM)

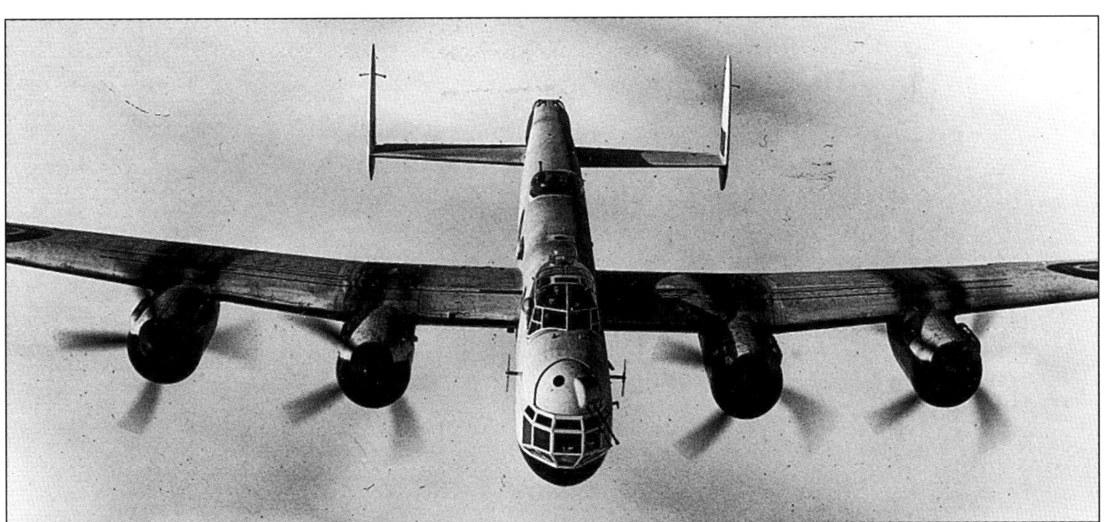

◀ 1944: DH 103 HORNET
The Hornet was a single-seat smaller version of the Mosquito and was intended for long-range fighter operations in the Pacific. It was the last piston-engined fighter to see RAF service and, with its two 2,070 hp Rolls-Royce Merlins, it was also one of the fastest. The RAF flew 211 Hornets, and the Royal Navy operated the Sea Hornet until 1956. Wing span 45 feet; maximum speed 472 mph. (RAFM)

◀ 1944: BRISTOL BRIGAND
A derivative of the Beaufighter, the Brigand was a light bomber powered by two Bristol Centaurus engines. The first 13 aircraft were delivered to the RAF as torpedo-bombers, but did not enter service. Brigand light bombers entered service in 1949, and 142 were built. The Brigand was the RAF's last piston-engined light bomber. Span 72 feet 4 inches; speed 360 mph. (IWM)

▶ 1945: BRISTOL TYPE 170
The Type 170 Bristol Freighter was a high-wing, short-range, general duty transport. Side-hinged nose doors gave unobstructed entry to the freight hold which was 32 feet long. The Wayfarer variant had a fixed nose and seats for 32 passengers. With two Bristol Hercules engines, Type 170s saw worldwide civil and military use. Span 98 feet; speed 240 mph. (BAe)

▼ 1945: HAWKER SEA FURY
The Sea Fury was a short-span Tempest II, lacking the latter's wing centre-section. Powered by a Bristol Centaurus engine, it entered service with the Fleet Air Arm in 1947. Sea Furies were operated in the ground support role during the Korean War and the type was exported to seven countries. Wing span 38 feet 5 inches; maximum speed (Mk 11) 460 mph at 18,000 feet. (IWM)

▲ 1945: AVRO TUDOR

The first British pressurised transport, the Tudor was designed as an interim airliner for the North Atlantic route pending the arrival of more modern aircraft such as the Comet. Early problems delayed production and, after two crashes, the type was withdrawn from service. A number of later models were produced with a longer fuselage and more powerful engines, and these were used successfully by charter companies. Wing span 120 feet; maximum speed 282 mph. (AMP)

▲ 1945: VICKERS VIKING

Combining the Wellington wing and the Warwick tail with a new stressed-skin fuselage, the Viking was a short-haul passenger aircraft designed for European routes. Powered by two 1,675 hp Bristol Hercules, the Viking carried 21 passengers and entered BEA service in September 1946. Later versions could carry 27 passengers. The Viking was exported to many countries and four were supplied to The King's Flight. Total production was 163 and one was used as a test-bed for the Nene jet engine. Wing span 89 feet 3 inches; maximum speed 263 mph at 10,000 feet. (AMP)

▲ 1945: DH 104 DOVE

The Dove was De Havilland's first post-war civil design and achieved worldwide success. A light transport offering seating for up to 11 passengers, the Dove was powered by two Gipsy Queen engines. Thirty were delivered to the RAF, where it was known as the Devon, and a number were used by the RN as the Sea Devon. DH 104 production continued for nealy 25 years and 542 were built. Wing span 57 feet; maximum speed 210 mph. (RAFM)

◀ 1946: DHC 1 CHIPMUNK

Designed by De Havilland of Canada, the tandem two-seat Chipmunk first flew in May 1946. Production totalled 218 built in Canada and 1,000 in England. The Chipmunk became the RAF's standard elementary trainer, replacing the Tiger Moth, and was exported to 14 overseas air forces. Large numbers of ex-RAF aircraft were civilianised and the Chipmunk is still in service with the RAF's Air Experience Flights. Wing span 34 feet 4 inches; maximum speed 138 mph. (RAFM)

◀ **1946: SUPERMARINE ATTACKER**
Supermarine's first jet design, the Attacker utilised the laminar-flow wing of the Spiteful. Powered by a 5,000 lb st Rolls-Royce Nene and with a conventional undercarriage, the Attacker was produced as a carrier-borne naval fighter. It entered service with No 800 Squadron in August 1951 and 149 were built. A further 36 Attackers were supplied to the Pakistan Air Force. Wing span 36 feet 11 inches; maximum speed 590 mph at sea level. (RAFM)

▶ **1947: ARMSTRONG WHITWORTH AW 52**
The AW 52, preceded by the experimental AW 52G glider, was built to explore the technology of tailless flight for a projected large airliner that would carry its passengers within the wing. Powered by two Rolls-Royce Nemo jet engines, two machines were built from one of which the pilot made the first successful ejection using a Martin Baker seat. The second AW 52 made only a few more flights before being scrapped. Wing span 90 feet; speed 500 mph. (IWM)

▲ **1947: AIRSPEED AMBASSADOR**
The Ambassador was designed in the later stages of the war but delays in production meant that only 23 aircraft were built. Twenty were operated by British European Airways, who named it the Elizabethan, and subsequently by a number of smaller airlines. Powered by two 2,700 hp Bristol Centaurus engines, the Ambassador could seat 47 passengers. Wing span 115 feet; maximum cruising speed 300 mph at 20,000 feet. (AMP)

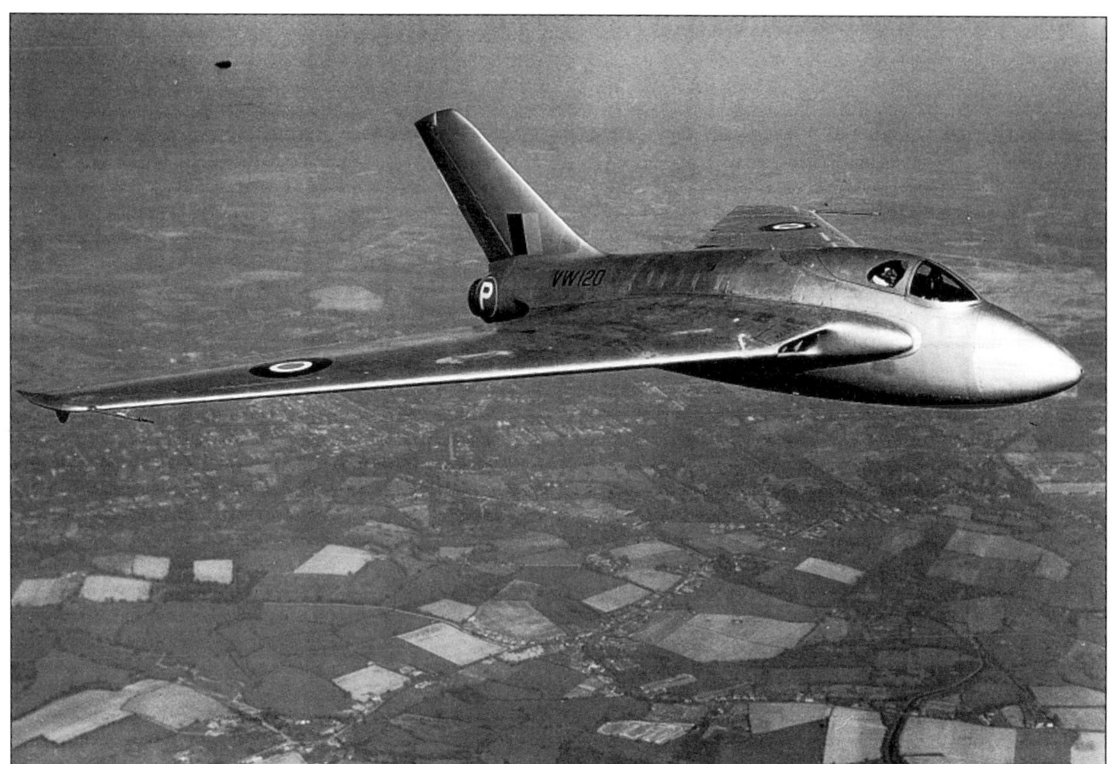

◀ **1948: DH108**

The DH108 was a single-seat research aircraft using a standard Vampire fuselage with wing and tail fin but no tailplane. Three were constructed to provide information on the behaviour of swept wings in connection with design work on the Comet airliner and the DH110 fighter. The third machine was the first British aircraft to exceed the speed of sound. Tragically, all three were lost with their pilots, who included the younger Geoffrey de Havilland. Wing span 39 feet; maximum speed 640 mph. (RAFM)

▲ **1948: HAWKER SEA HAWK**

Hawker's first jet aircraft, the Sea Hawk was conceived in 1944 as a carrier-borne fighter. Powered by a Rolls-Royce Nene turbojet, it entered service with No 806 Squadron in March 1953. Initially built at Hawker's Kingston factory, development and production were subsequently transferred to Armstrong Whitworth at Coventry. More than 430 Sea Hawks were built for the Royal Navy and the type also served in substantial numbers with the West German, Dutch and Indian navies. Wing span 39 feet; maximum speed Mach 0.84 at 36,000 feet. (GB)

▲ 1948: VICKERS VISCOUNT

The first turbine-powered passenger airliner, the Viscount was bought new by over 60 operators in some 40 countries. First flown on 16 July 1948, a total of 445 was built. Powered by four 1,380 hp Rolls-Royce Darts, it could carry 32 passengers although, later, more powerful marks could seat up to 65 passengers. One aircraft was fitted with two Rolls-Royce Tay turbojets and was later used for powered flying control trials. Wing span 93 feet 9 inches; maximum speed (Type 810) 357 mph at 20,000 feet. (AMP)

▼ 1949: BRISTOL BRABAZON

Developed from ideas for a large heavy bomber the Brabazon was a 100-seat passenger airliner with a range of 5,500 miles. Powered by eight Bristol Centaurus driving four pairs of contra-rotating propellers, it first flew on 4 September 1949. The Brabazon was ahead of its time but the project was cancelled for political and financial reasons. Wing span 230 feet; speed 300 mph. (IWM)

◀ 1948: PERCIVAL PRINCE

A high-wing monoplane powered by two Alvis Leonides radial engines, the Prince was an eight-passenger feeder-liner, executive transport and survey aircraft. Some were supplied to the Royal Navy as the Sea Prince. A larger version was known as the President and the RAF also operated a number of these as Pembrokes. Wing span (Prince) 56 feet; speed 229 mph. (RAFM)

▲ ▼ 1949: ENGLISH ELECTRIC CANBERRA

Powered by two Roll-Royce Avon jet engines, the Canberra was the RAF's first jet bomber. The basic design was flexible enough for the aircraft to be developed as a photo-reconnaissance (PR9, below), night intruder, bomber, (B2T, above), trainer and ECM aircraft. First flown in May 1949, the Canberra was an outstanding success and 1,376 were built, including over 400 produced in the USA. It has been operated by many overseas air forces and is still in RAF service in 1995. Wing span 64 feet; maximum speed 518 mph at sea level. (RB) (BAe)

◀ 1949: ARMSTRONG WHITWORTH AW55 APOLLO
Designed to carry up to 30 passengers over 1,000 miles at a cruising speed of 300 mph, the Apollo was a handsome aircraft powered by four Armstrong Siddeley Mamba propeller-turbines. These engines proved unreliable and only two Apollos were built, one subsequently being used by the Empire Test Pilot's School. Wing span 92 feet; maximum speed 330 mph. (RAFM)

▲ 1949: AVRO SHACKLETON
The Shackleton was a direct descendant of the Lancaster and Lincoln. Powered by four Rolls-Royce Griffon engines, it was a longe-range maritime patrol aircraft and entered RAF service in 1951. The Mk 2 had a new streamlined nose and the Mk 3 was a tricycle undercarriage version. Modified Mk 2s were used as airborne early warning aircraft right up to 1991. Wing span 120 feet speed 302 mph. (GB)

▲ **1949: VICKERS VALLETTA/VARSITY**

The Valletta was a military transport version of the Viking airliner and was supplied to RAF Transport Command. It had a strengthened floor and a large loading door in the left-hand side of the fuselage. It could carry 34 fully-equipped troops or 20 stretcher cases. Some were used for navigation and radar training. The Varsity was a multi-engined pilot trainer which was also used for training other aircrew. It was distinguishable by its tricycle undercarriage. Production of both types amounted to 426. Wing span (Varsity) 95 feet 7 inches; maximum speed (Varsity) 288 mph. (GB)

▲ **1949: AVRO 707**

Designed to provide delta-wing data for the Vulcan bomber project, the Avro 707 was one-third the size of the production aircraft. With its single Rolls-Royce Derwent turbojet engine, both the 707 and its successor, the 707B, had a dorsal engine air intake and were used specifically for low-speed trials. Two 707As were built to research the high-speed characteristics of the wing, and these had wing-root intakes. The final, 707C, version was a two-seater which also had wing-root intakes. Wing span (707A/C) 34 feet 2 inches. (SSPL)

◀ **1949: DH COMET**

Britains's first pure-jet airliner made its maiden flight on 27 July 1949. Powered by four DH Ghost turbojets, the Comet 1 entered service with BOAC in April 1952, introducing the world's first jet airliner service on 2 May with a flight to Johannesburg. Structural problems grounded the Mk 1s, but 15 Comet 2s were delivered to the RAF, and the larger Comet 4 was sold to a number of airlines, including BOAC and BEA. Wing span 118 feet; cruising speed 532 mph (details for Comet 4B). (BAe)

▲ **1950: DH114 HERON**

The Heron was a four-engined version of the Dove and was designed for short and medium-haul operations in areas where there were no proper aerodromes. The Mk 1 had a fixed undercarriage but the Mk 2 had a retractable undercarriage and was 20 mph faster. Although only 143 were built, the Heron saw service in 30 countries as a 17-seater airliner or luxury private transport. Three were delivered to The Queen's Flight. Wing span 71 feet 6 inches; cruising speed 183 mph. (AMP)

▲ **1950: PERCIVAL PROVOST**

The Provost succeeded the Prentice in 1953 as the RAF's standard basic trainer. Powered by a 1,550 hp Alvis Leonides engine, the Provost prepared pupil pilots for advanced training on the jet Vampire. Production totalled 397 and the Provost was used until the end of 1961, although some were retained until the end of 1969. The Provost combined high performance with exceptional manoeuvrability. Wing span 35 feet 2 inches; maximum speed 200 mph. (GB)

▲ **1950: BLACKBURN BEVERLEY**

This four-engined medium-range tactical transport entered RAF service in 1956 and served all over the world until 1968. Its capacious hold could carry as many as nine jeeps or a single bulldozer, and there was room for 36 passengers in the tail boom. Two prototypes and 47 production aircraft were built. Wing span 162 feet; maximum speed 238 mph. (RAFM)

▲ 1951: HAWKER HUNTER
Nearly 2,000 Hunters in many variants were built and the aircraft has seen worldwide service. A single-seat fighter, the Hunter was powered by a Rolls-Royce Avon turbojet, although one version was fitted with the Armstrong Whitworth Sapphire. Hunters entered RAF service in July 1954 with the Air Fighting Development Squadron and then with No 43 Squadron. The aircraft was subsequently developed as a ground attack fighter and was latterly used for training. A two-seat version was built and the Hunter was also used by the Royal Navy. Wing span 33 feet 8 inches; maximum speed Mach 0.95 at 36,000 feet. (BAe)

◀ 1951: VICKERS VALIANT
The Valiant was the first of the RAF's four-jet V-bombers and first flew on 18 May 1951. Powered by four 10,050lb-thrust Rolls-Royce Avon turbojets, the Valiant could carry 21,000lb of nuclear or conventional weapons. Entering service with No 138 Squadron in January 1955, the Valiants were also operated as strategic reconnaissance aircraft and as in-flight refuelling tankers. The first British atomic and hydrogen bombs were dropped from Valiants. Wing span 114 feet 4 inches; maximum speed 567 mph at 30,000 feet. (RAFM)

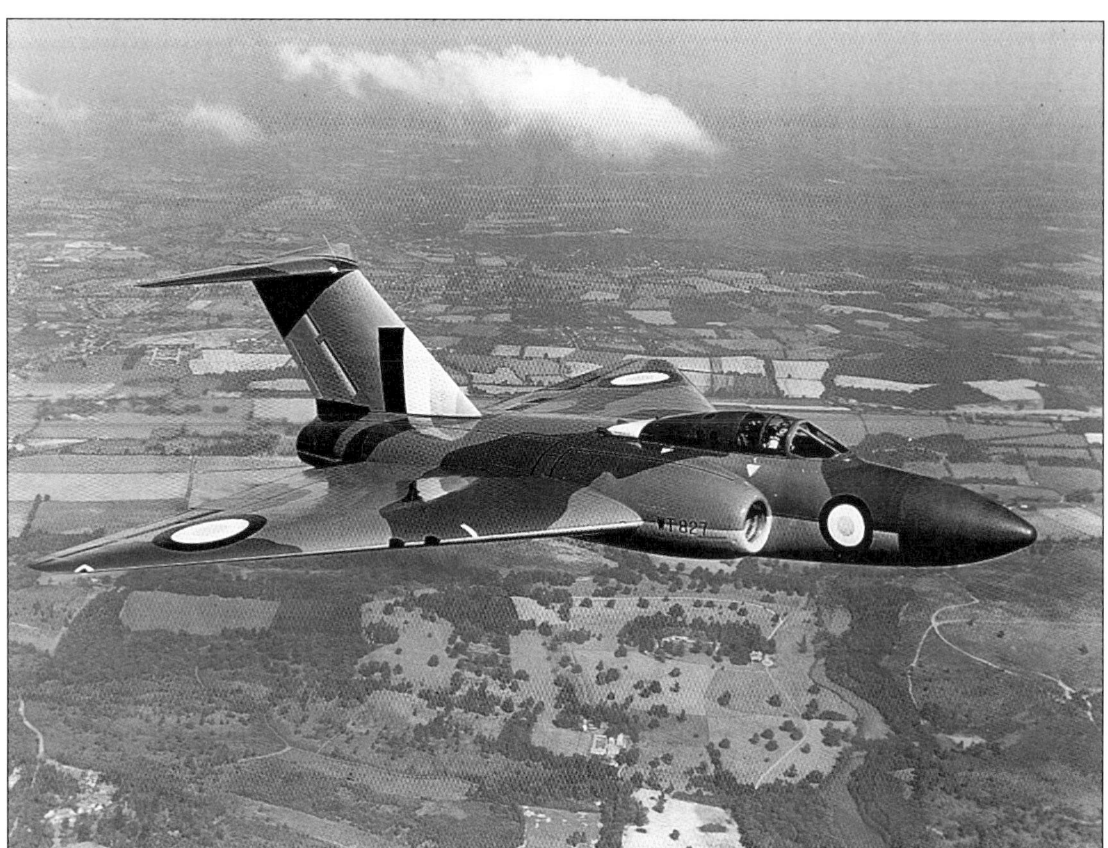

◀ 1951: GLOSTER JAVELIN
The Javelin was a large two-seat delta wing all-weather fighter powered by two Armstrong Siddeley Sapphire turbojets. The Mk1 entered service with No 46 Squadron in February 1956. Developed through several variants to the Mk9, 302 Javelins were built in all and the last was retired from front-line service in June 1967. The aircraft carried four Aden cannon and the later marks could carry four Firestreak air-to-air missiles. Wing span 52 feet; maximum speed (Mk9) 610 kts at sea level. (RAFM)

▲ 1952: SUPERMARINE SWIFT
The Swift was first produced as a fighter for the RAF. With a fuselage not unlike the Attacker's but having swept wings and a tricycle undercarriage, it was powered by a Rolls-Royce Avon jet engine. Protracted development led to its role being changed to tactical reconnaissance and, as the FR5, the Swift entered service with No 2 Squadron in 1956 and subsequently with No 79 Squadron. Wing span 36 feet; maximum speed 700 mph at sea level. (GB)

1952: AVRO VULCAN
Ranking alongside the Lancaster in the annals of Avro's fame, the Vulcan was the world's first delta-wing jet bomber. The B1 had a straight leading edge, (as on the prototype, left) and entered RAF service in 1956. It provided part of the UK's nuclear deterrent, until that role was taken over by the Royal Navy in 1969, and the B2 also carried the Skybolt missile (lower left). The Vulcan was then switched to a conventional role, operating latterly at low level. Some aircraft had a maritime reconnaissance role and a few were modified as in-flight refuelling tankers. Withdrawn from service at the end of 1982, one aircraft was kept airworthy for aerial displays for a number of years. Wing span (B2) 111 feet; speed (B2) 645 mph. (SSPL) (BAe)

▲ 1952: BRISTOL BRITANNIA

The Britannia was designed in response to BOAC's 1946 requirement for a Medium Range Empire transport. Powered by four 3,780 hp Bristol Proteus turboprops, the Britannia could carry up to 139 passengers for over 4,000 miles. It was operated by BOAC, the RAF and a number of international airlines before having a second life with charter companies. Eighty-five Britannias were built, production ceasing earlier than hoped because of the imminence of the first jet airliners. Three Canadian versions were produced: the Argus maritime partol aircraft, the Yukon military transport, and the civil CL44 freighter. Wing span 142 feet 3 inches; cruising speed 357 mph. (BAe)

◀ 1953: ENGLISH ELECTRIC LIGHTNING

First flown on 4 August 1954 as the P.1A, the Lightning had an exceptional performance. Powered by two Rolls-Royce Avons, it was able to fly at more than twice the speed of sound and could climb to 40,000 feet in two and a half minutes. The last RAF Lightnings were withdrawn from operational service in 1988. Variants of the Lightning were exported to Saudia Arabia and Kuwait. Wing span 34 feet 10 inches; maximum speed at high altitude Mach 2.2. (BAe)

1954: PERCIVAL JET PROVOST

The Jet Provost basic trainer replaced the piston-engined Provost and introduced all-through jet training to the RAF. Although all marks of the Jet Provost used the Viper engine, the T1 (upper left) had the earliest version of the engine and was quite underpowered. The T3 (lower left) was the main mark used, and 201 were delivered. The later Jet Provost T4, of which 185 were built, had an uprated Viper engine that provided 40% more thrust than that available on the T3. The final version of the Jet Provost was the T5 (below) which had a pressurised cockpit. T5 production totalled 110. Wing span 35 feet 4 inches; maximum speed (T5) 440 mph at 25,000 feet.

(RAFM) (BAe)

▲ 1956: SUPERMARINE SCIMITAR
The Scimitar was a single-seat naval jet fighter powered by two 10,000 lb st Rolls-Royce Avon engines. First flown in January 1956, it featured power controls and blown flaps. The Scimitar entered operational service with No 803 Squadron in June 1958 and embarked on HMS Victorious in September. Despite having four 30mm Aden cannon, the Scimitar was operated mainly in the low-level strike role. Span 37 feet 2 inches; speed Mach 0.968 at sea level. (GB)

◀ 1956: SCOTTISH AVIATION TWIN PIONEER
Following the success of the Pioneer, a twin-engined version was developed which could seat up to 16 passengers and two crew. Eighty-five were built, of which 40 were supplied to the RAF and the remainder sold worldwide. Powered by two Alvis Leonides engines, the Twin Pioneer required an area only 900 feet by 100 feet from which to operate. Wing span 76 feet 6 inches; maximum speed 165 mph. (GB)

◀ 1957: DH 110 SEA VIXEN
The final development in the Vampire/Venom sequence, the DH110 evolved into a carrier-borne all-weather fighter which entered RN service in 1957 as the Sea Vixen. One hundred and nineteen Mk1s were built and were followed by 29 Mk2s. An underwing refuelling pack enabled one Sea Vixen to refuel another in mid-air. Wing span 50 feet; maximum speed 645 mph at 10,000 feet. (RAFM)

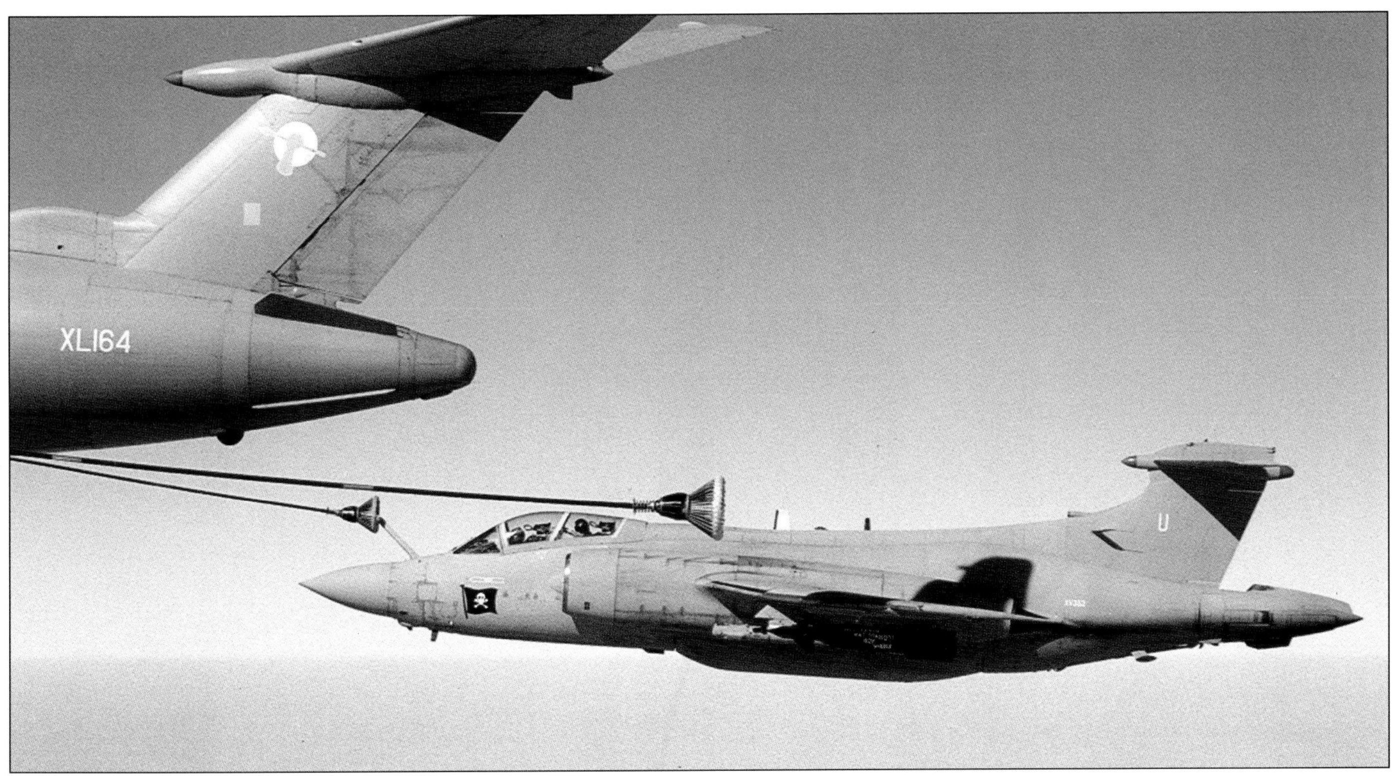

▲ 1958: BLACKBURN BUCCANEER

The Buccaneer was an advanced, high-speed, low-level, strike aircraft designed for carrier operations. The first prototype flew in 1958 and Mk1s entered Royal Navy service in 1962. The Mk2 was powered by the Rolls-Royce Spey engine and in this form served with the Royal Navy until 1978. Many ex-Navy aircraft were passed to the RAF who had begun operating the type in 1969. The last Buccaneers were retired in 1994. Wing span (Mk2) 44 feet; maximum speed 645 mph at sea level. (BAe)

◀ 1959: ARMSTRONG WHITWORTH ARGOSY

The Argosy was the last aircraft to bear the Armstrong Whitworth name and was originally designed as a military medium-range freighter to carry scout cars and other light vehicles for the Army. Powered by four Rolls-Royce Dart turbo-propeller engines, the Argosy could carry 31,000lb of cargo which could be loaded through nose and tail cargo doors. The Argosy was used by Riddle Airways, British European Airways and, later, a number of charter companies. Fifty-six were supplied to the RAF. Wing span 115 feet; cruising speed 285 mph.

(BAe)

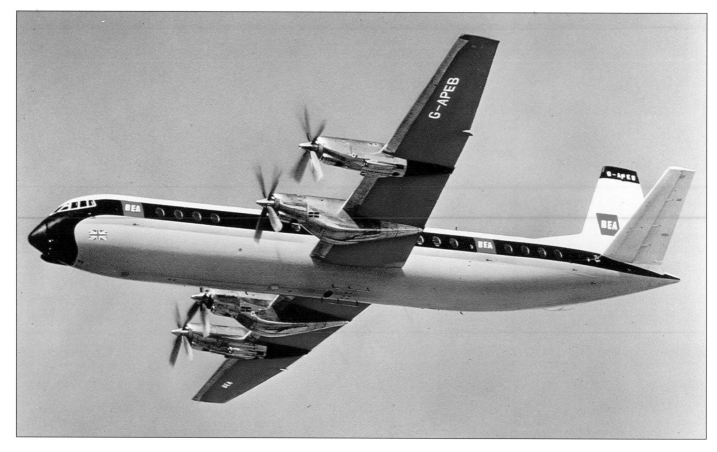

▲ 1959: VICKERS VANGUARD

The Vanguard was designed as a successor to the Viscount. Powered by four 5,545 ehp Rolls-Royce Tyne turboprops, it could carry 139 passengers and was ordered by British European Airways and Trans-Canada Air Lines. The advent of the jet airliner meant that only 44 Vanguards were built and many of these were subsequently converted to freighters. Wing span 118 feet; speed 400 mph. (AMP)

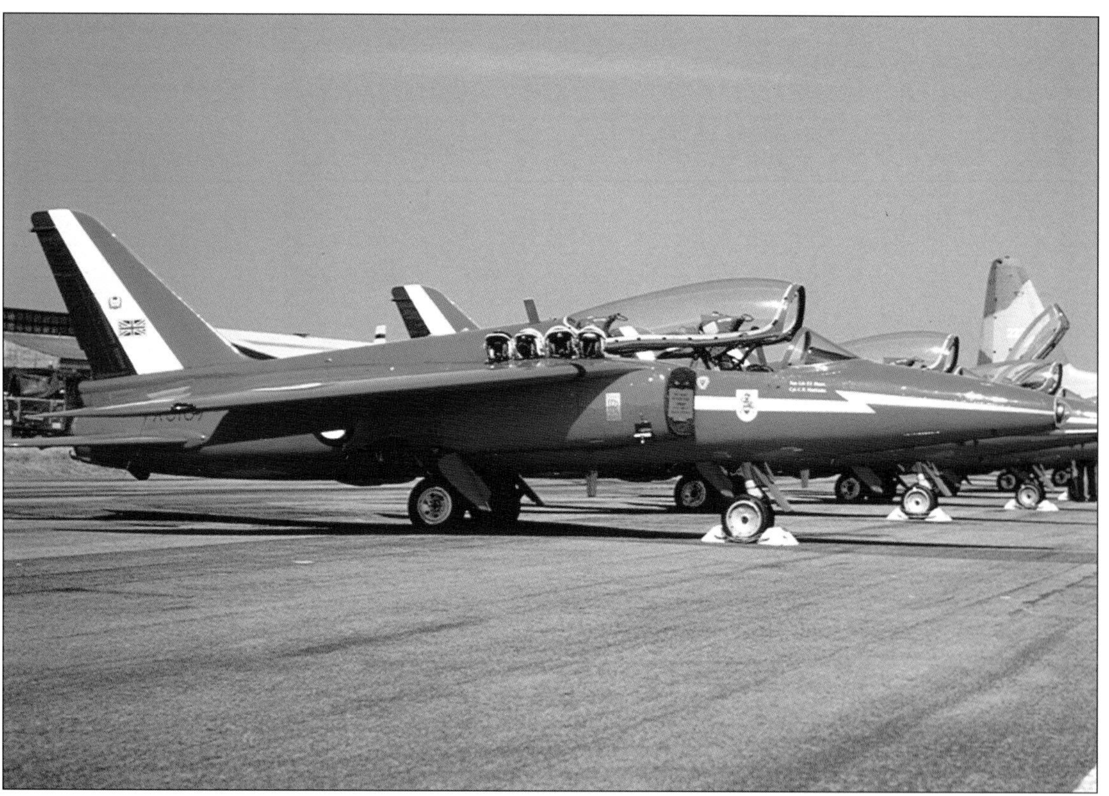

◀ 1959: FOLLAND GNAT

Originally developed as a small, light fighter, the Gnat was put into production as an advanced trainer for the RAF. Powered by a 4,230lb thrust Bristol Siddeley Orpheus jet engine, the Gnat's two seats were in tandem and the aircraft was capable of supersonic flight in a shallow dive. It entered RAF service in 1962 and 105 were built. Gnats were subsequently used by the Red Arrows Aerobatic Team. The original single-seat fighter version was exported to India and Finland. Span 24 feet; speed Mach 0.97. (GB)

◀ ▲ 1960: HS 748

Beginning as a private venture, the 48-seat Avro - subseqently Hawker Siddeley - 748 airliner has been a most successful export. Over 370 aircraft of all marks have been built in the UK with a further 89 assembled in India. In RAF service, the aircraft was named Andover, the C1 being a specially lengthened version with a rear loading door and the CC2 being the 748 Series 2. This latter mark was also operated by The Queen's Flight. The 748 was the last type to carry the name of Avro. Wing span 98 feet 6 inches; cruising speed 284 knots.

(BAe) (GB)

▲ 1962: VICKERS VC10

Powered by four rear-mounted Rolls-Royce Conway jets, the VC10 first flew on 29 June 1962. In its original form it could carry up to 135 passengers but the longer Super VC10 could seat 163. VC10s were ordered by BOAC and a number of overseas airlines as well as by RAF Transport Command. More recently, the RAF bought a number of ex-civil aircraft and converted them into air-to-air refuelling tankers. Wing span 146 feet 2 inches; maximum speed 500 mph. (RB)

▲ 1962: HAWKER SIDDELEY TRIDENT

Built in response to BEA's request for a 600 mph short-haul jet airliner, the Trident went straight into production off the drawing board. Powered by three Rolls-Royce Speys grouped at the tail, the Trident 1 could carry up to 97 passengers and was the world's first airliner fitted with automatic landing facilities. The aircraft was developed through several versions culminating in the much larger Trident 3 seating up to 179 passengers. Over 100 Tridents of all marks were built, some of which were bought by the Chinese national airline. Wing span 98 feet; maximum cruising speed 550 mph (details for Trident 3B). (AMP)

1962: HS125

A jet replacement for the Dove executive transport this De Havilland aircraft first flew in August 1962 and has been a great success. Subsequently developed by the Hawker Siddeley Group, into which De Havilland was absorbed, over five hundred and fifty 125s of all marks have been sold worldwide. The RAF use the 125 for VIP duties and, as the Dominie T1, also for navigator training (upper left). The -800 derivative of the 125 (top) is known as the Hawker 800 and is produced by Raytheon in America. The 125-1000 is the latest member of the 125 family (lower left). Wing span 51 feet 4 inches; maximum cruising speed 525 mph.

(BAe)

◀ 1963: BAC ONE-ELEVEN
Developed from a design from Hunting Aircraft, the BAC 1-11 became one of Britain's best-selling airliners. The first prototype flew in August 1963 and the first customer was British United Airways. Since then, the aircraft has been developed through several variants and sold and operated worldwide. It has also been built under licence in Romania. Biggest of the variants was the Series 500 which could seat 119 passengers. Wing span 93 feet 6 inches; maximum cruising speed 541 mph at 21,000 feet. (Details for Series 500). (BAE)

▲ 1964: BRITISH AIRCRAFT CORPORATION TSR 2
Standing for Tactical, Strike and Reconnaissance, the TSR.2 was designed as a replacement for the Canberra. Powered by two Bristol Siddeley Olympus jet engines, it featured a small delta-wing with anhedral tips mounted high on a long fuselage. It first flew on 27 September 1964 and early flight trials were highly promising. A second aircraft was completed but had not flown when a newly-elected Labour government cancelled the project on 6 April 1965. Wing span 37 feet 2 inches; maximum speed at high altitude Mach 2 plus. (BAE)

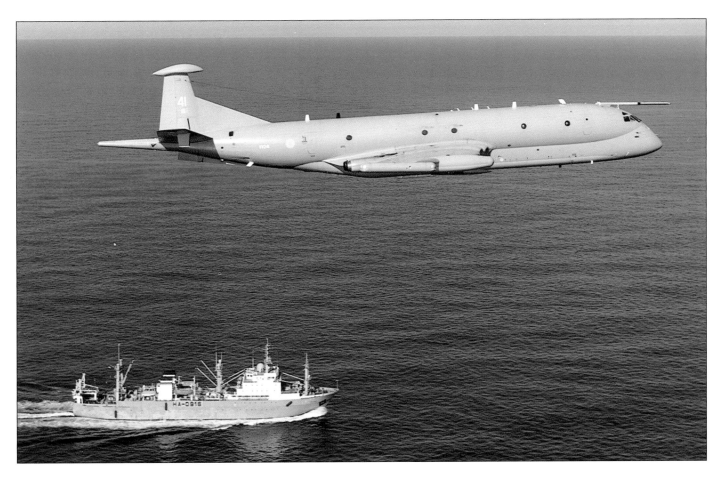

▲ 1967: HAWKER SIDDELEY NIMROD

The Nimrod maritime reconnaissance and anti-submarine aircraft was developed from the Comet airliner. Powered by four Rolls-Royce Spey engines, each producing 12,160 lb thrust, the Nimrod can fly at high speed to its patrol area and then loiter on two engines. With a crew of 13, the aircraft can carry torpedoes, bombs and Harpoon air-to-surface missiles in the internal weapons bay, and Sidewinder air-to-air missiles externally. Forty-six were built including three specially modified R1 variants for electronic intelligence gathering. Wing span 114 feet 10 inches; maximum speed 575 mph. (BAE)

◀ 1967: BAC STRIKEMASTER

The Strikemaster is a development of the Jet Provost T.5 trainer and has been exported to a number of air forces as a tactical ground-attack aircraft. Powered by a 3,410lb thrust Rolls-Royce Viper, it has a pressurised cockpit and can carry up to 3,000lb of bombs and rockets. Wing span 36 feet 10 inches; maximum speed 472 mph at 20,000 feet. (BAE)

1966: HAWKER SIDDELEY HARRIER

The Harrier, developed from the P1127 Kestrel (top left), was the world's first vertical/short take-off and landing (V/STOL) jet fighter and entered RAF service with No 1 Squadron on 1 April 1969. Powered by a single 21,500lb Rolls-Royce Bristol Pegasus vectored-thrust turbofan, the Harrier has been considerably developed since the first GR1 variant. GR3, T4 and GR5 versions have all seen RAF service, and the GR7 and T10 are now on the front line. The Sea Harrier entered service with the Royal Navy in 1979. Harriers are also in service with the United States Marine Corps and the Spanish and Italian forces. RN and RAF Harriers played a large part in the 1982 Falklands conflict. Wing span 25 feet 2 inches; maximum speed 740 mph at sea level. (GR3). (BAe)

▲ Royal Navy variants of the Harrier: From back to front - T4N, FRS1, F/A2

◀ AV-8A Matador of the Spanish Navy

▲ Harrier T10 entered RAF service in 1994

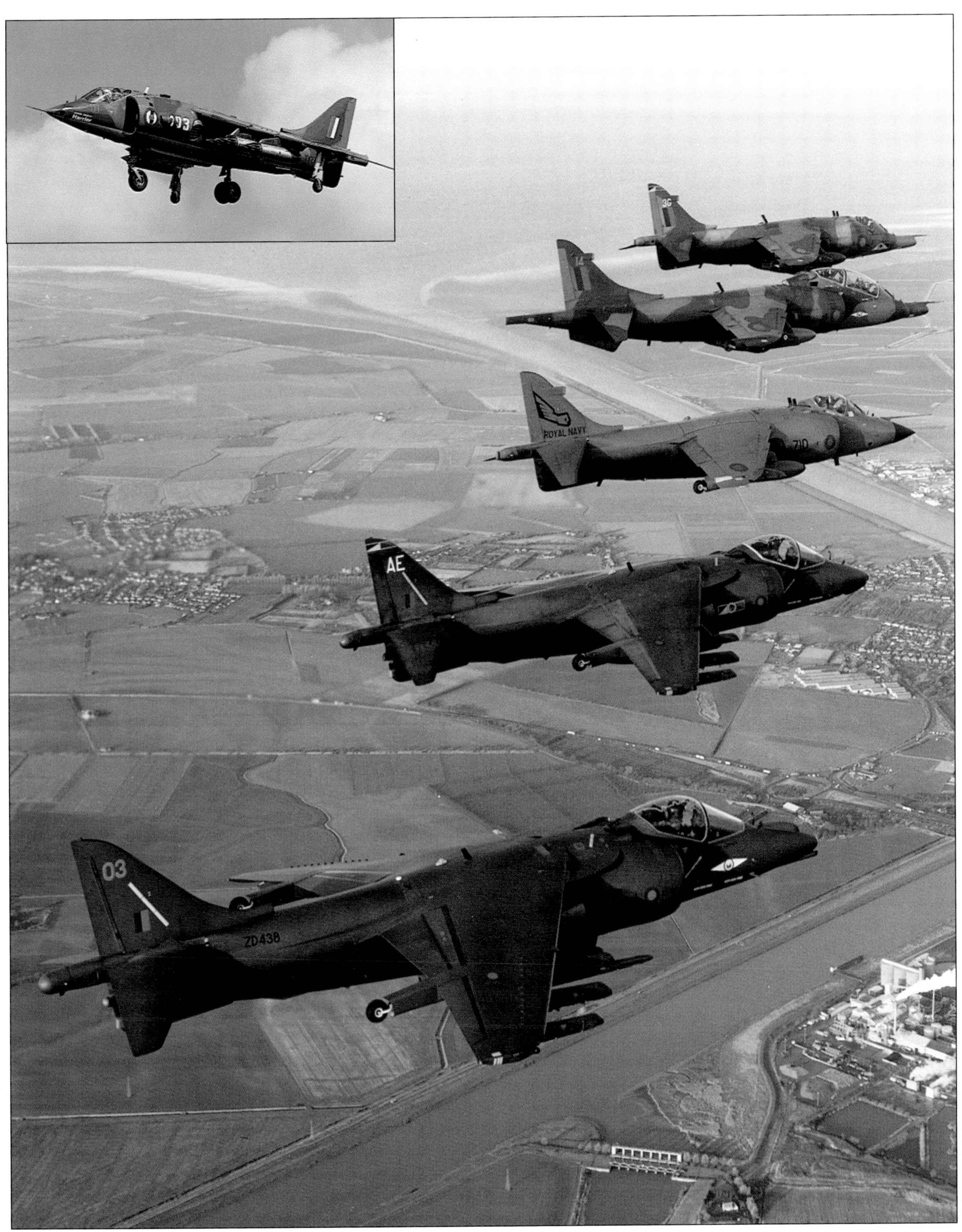

▲ Most of the Harrier family: The GR1 (inset) and, from back to front, the GR3,T4, FRS1, GR5 and GR7. (BAe)

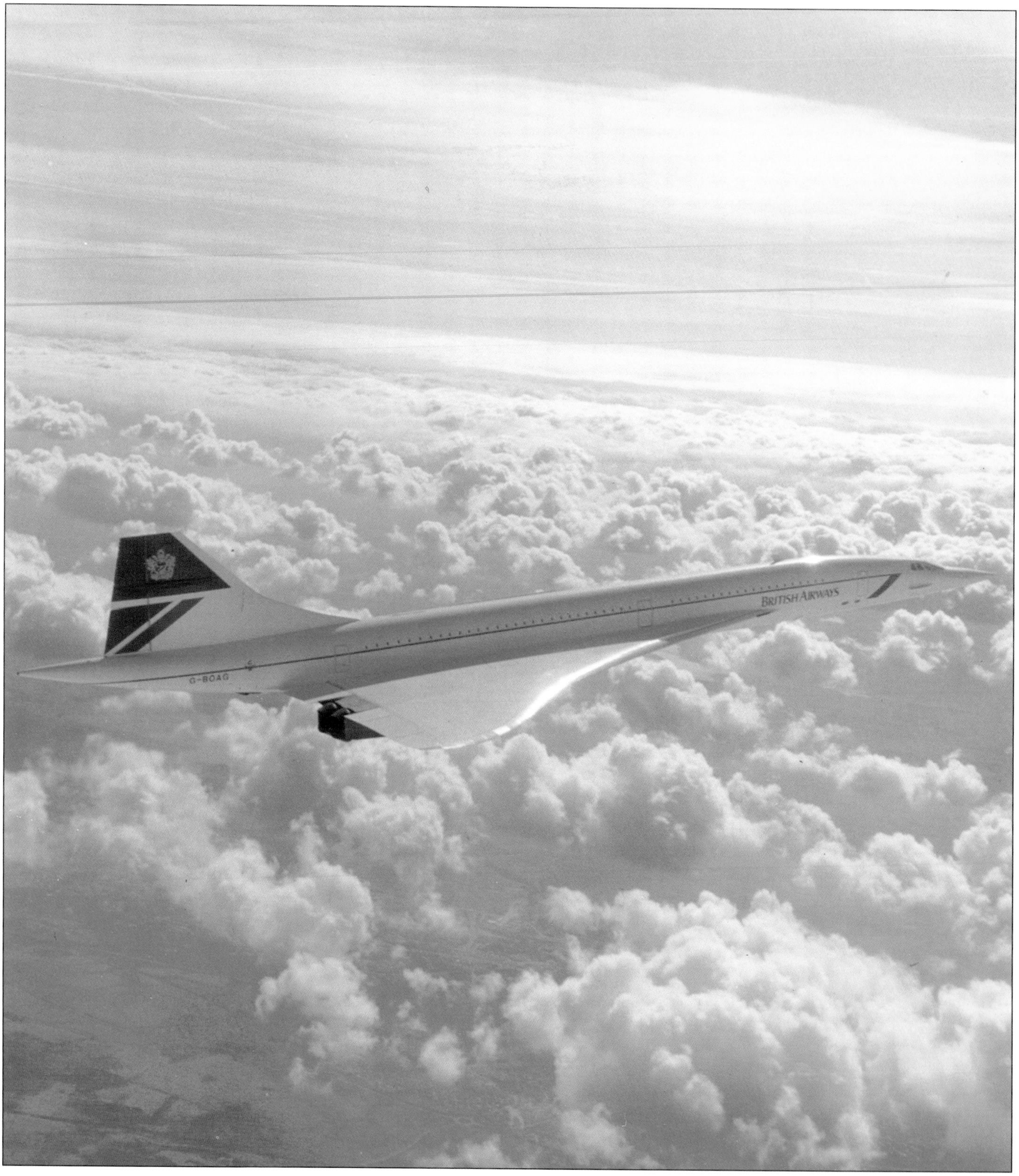

▲ **1967: BAC/AEROSPATIALE CONCORDE**

The Anglo-French Concorde is the only successful supersonic airliner. The prototype flew on 11 December 1967 and Concorde entered service with British Airways and Air France on 21 January 1976. Powered by four 38,050lb thrust Rolls-Royce/SNECMA Olympus turbojets, it has a range of over 4,000 miles and can fly from London to New York in about 3.5 hours. British Airways currently operates seven Concordes and Air France six, all aircraft carrying 100 passengers. Wing span 83 feet 10 inches; maximum cruising speed Mach 2.04 at 51,000 feet. (AMP)

◀ **1967: SCOTTISH AVIATION JETSTREAM**
Originally a Handley-Page design, the Jetstream is a passenger feeder liner and executive transport. First models were powered by two Turbomeca Astazou turboprops but more recent aircraft have two Garrett turboprops. Twenty-six Jetstreams were supplied to the RAF to replace the Varsity. Latest variants are the stretched Jetstream 41 and the much larger Jetstream 61, the latter being a development of the Avro 748. Wing span (Jetstream 41) 60 feet 5 inches; maximum cruising speed (Jetstream 41) 340 mph. (BAe)

▲▶ **1969: SEPECAT JAGUAR**
A collaborative project between the British Aircraft Corporation and Dassault/Breguet, the Jaguar has been produced both as a single-seat tactical strike-fighter and two-seat advanced trainer. One hundred and sixty five of the former and 35 of the latter have been supplied to the RAF. The French Air Force operate a different version. Jaguars based on the RAF's model have been exported to India (photo above), Ecuador, Oman and Nigeria. The aircraft entered RAF service in April 1975 and still equips three squadrons in 1995, some having been upgraded to carry a Thermal Imaging Laser Designator pod. Wing span 28 feet 6 inches; maximum speed Mach 1.6 at 36,000 feet. (BAe)

◀ 1971: SCOTTISH AVIATION BULLDOG

Originally designed by Beagle Aircraft, production of the Bulldog was taken over by Scottish Aviation which built 130 for the RAF. A side-by-side two-seat primary trainer, it entered service in April 1973 as a replacement for the Chipmunk in Flying Training Schools and University Air Squadrons. Bulldogs have also been exported to a number of air forces. Wing span 33 feet; maximum speed 150 mph. (BAe)

▲ 1972: AIRBUS A300

The Airbus is a European collaborative project and is proving a successful challenge to the previous dominance by the Americans of civil jet airliner manufacture. British Aerospace has a 20% stake in the project. A number of variants of the original 252-seat A300 have been produced, the largest and most recent being the 440-seat A340, which has the longest range of any airliner. The Airbus is in service with airlines around the world. The photo above shows an A320 in Air Canada colours, an A300-600R in Korean Air livery, an A310 in Air France colours, and an A340 in the livery of Airbus Industrie. Wing span (A340) 197 feet 8 inches; maximum operating speed (A340) Mach 0.86. (BAe)

1974: HAWKER SIDDELEY HAWK

The Hawk is a tandem two-seat advanced jet trainer which replaced the Gnat and the Hunter in RAF service. One hundred and seventy five were ordered off the drawing board and the first were delivered to No 4 FTS in November 1976. Hawks are flown by the Red Arrows, the RAF's Aerobatic Team, and the type has been exported to no less than 20 countries. A single-seat combat version - the Hawk 200 - has also been built. Wing span 30 feet 10 inches; maximum speed 615 mph at sea level.

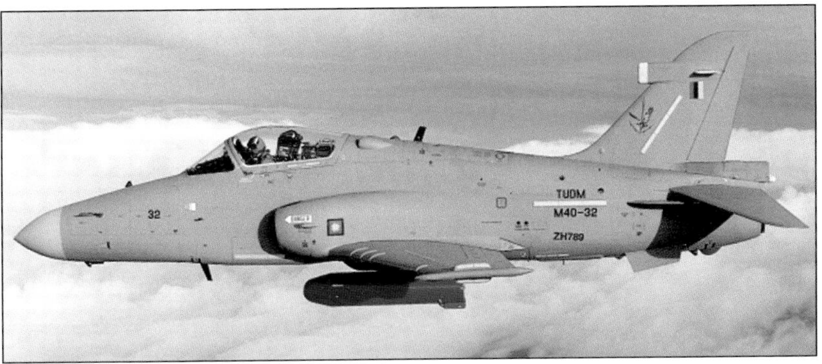

The Hawks shown here are (clockwise from bottom left): A Mk 63 from Abu Dhabi, a Swiss Mk 66, a Mk 60 from Zimbabwe, TMk1s of the RAF, T-45A Goshawks of the US Navy, and a Malaysian Air Force Hawk Mk 208. (BAe)

1974: PANAVIA TORNADO

Originally known as the MRCA (Multi-Role Combat Aircraft), the Tornado is an Anglo-German-Italian collaborative project. First flown on 14 April 1974, the Tornado IDS (Interdictor/Strike) version is in service with the RAF as the GR1, the GR1A (left) being a reconnaissance version and the GR1B a maritime attack variant. GR1s receiving a mid-life update will receive the designation GR4. The Tornado ADV (Air Defence Variant) is a long-range fighter in service with the RAF as the F3 (bottom). Tornados are also operated by the German, Italian and Saudi Arabian forces. Wing span (unswept) 45 feet 8 inches; speed (F3) 1,453 mph at 36,000 feet. (BAe)

◀ **1979: BRITISH AEROSPACE 146 REGIONAL JETLINER**
First flown in September 1981, the BAe 146 is a high wing airliner which is very quiet and able to operate from short strips. Able to carry up to 128 passengers, the 146 is in worldwide service and is also operated by No 32 (Royal) Squadron RAF. Production ceased after 222 aircraft in three versions had been built. The range has now been replaced by the Regional Jetliner family of four variants, which have substantially updated interiors and Textron Lycoming LF507 turbofans. Wing span 86 feet; maximum operating speed 351 mph. (BAe)

▲ **1994: EUROFIGHTER 2000**
A consortium of Britain, Germany, Italy and Spain was formed to produce this single-seat agile fighter. The aircraft is of canard delta-wing shape and employs fly-by-wire controls, advanced avionics and multi-function cockpit displays. Construction is of carbon-fibre composites, aluminium, lithium alloy and titanium. Eurofighter 2000 first flew on 27 March 1994 after the successful Experimental Aircraft Project, EAP, and will equip nine RAF squadrons early in the 21st century. Span 34 feet 6 inches; speed 1,321 mph. (BAe)

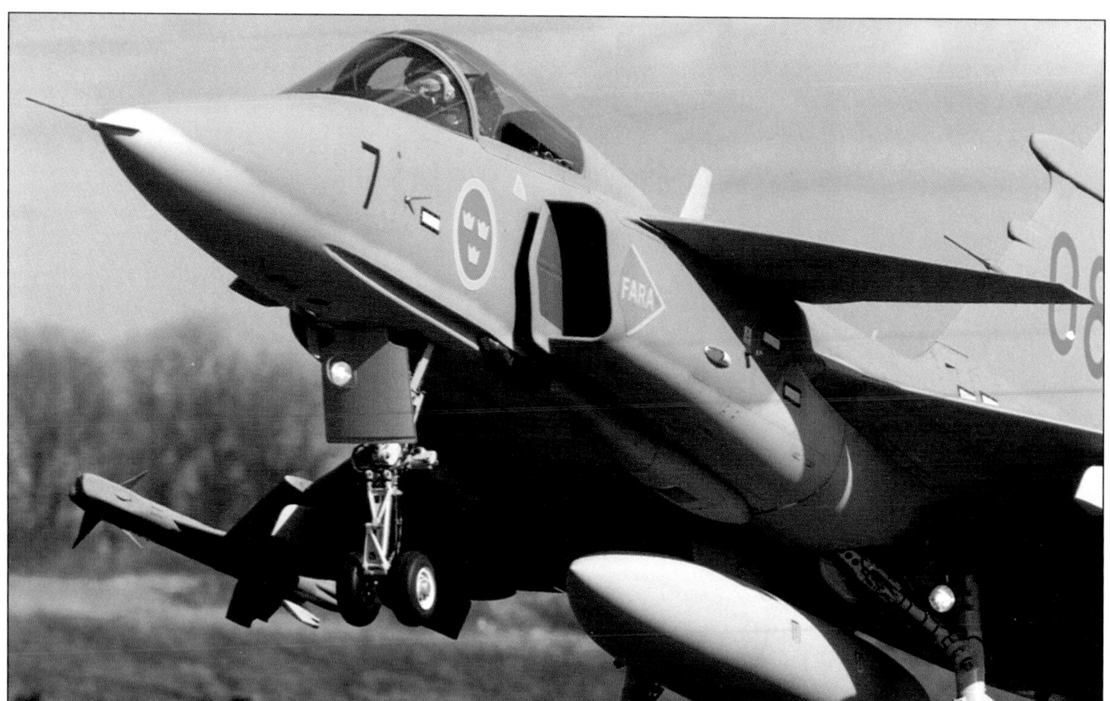

◀ 1995: SAAB GRIPEN
This new-generation multi-role fighter aircraft is an exciting addition to the British Aerospace product range. A joint venture agreement with Saab means that Gripen has the potential to be a great export success, especially with the worldwide marketing and support experience that BAe can bring to bear. Early wing sets for the Gripen project were built at Brough where the collaboration with Saab Military Aircraft will continue. (BAe)

▶ 2000+: AIRBUS INDUSTRIE A3XX
As a partner in Airbus Industrie, BAe is involved in the study for a very high capacity aircraft. This study runs in parallel to the Very Large Commercial Transport (VLCT) study which is investigating the feasibility of a 600-800 seat airliner. The artist's impression shows a double-decker concept. (BAe)

◀ 2000+: ASTOVL PROJECT
An exclusive partnership has been agreed with McDonnell Douglas to develop an Advanced Short Take-off / Vertical Landing aircraft for the 21st century. The BAe / McDonnell Douglas team is already at the forefront of VSTOL technology with the Harrier II Plus joint project. (BAe)